THE YOHKOH (SOLAR-A) MISSION

Library of Congress Cataloging-in-Publication Data

ISBN 978-94-010-5161-3   ISBN 978-94-011-2626-7 (eBook)
DOI 10.1007/978-94-011-2626-7

Printed on acid-free paper

THE YOHKOH (SOLAR-A) MISSION

A Series of Invited Contributions

Edited by

ZDENĚK ŠVESTKA

Laboratory for Space Research Utrecht, SRON, The Netherlands
and
Center for Astrophysics and Space Sciences, UCSD, La Jolla, CA 92093, U.S.A.

and

YUTAKA UCHIDA

Department of Astronomy, University of Tokyo, Bunkyo-ku, Tokyo 113, Japan

Reprinted from Solar Physics, Volume 136, No. 1, 1991

SPRINGER SCIENCE+BUSINESS MEDIA, B.V.

KATSUO TANAKA

(1943–1990)

On January 2, 1990, our colleague, the prominent solar physicist Katsuo Tanaka died in Tokyo after a long battle with leukemia. He was only 46.

Born in Tokyo in 1943, Tanaka graduated from the University of Tokyo in 1965. He was awarded a Ph.D. in 1971 for his study of the solar chromosphere based on the eclipse data.

In 1970 he began working with the Tokyo Astronomical Observatory, where his interests shifted to the observation of solar activity, particularly solar flares, in high resolution. In 1971, he began a cooperative research project with H. Zirin at the Big Bear Solar Observatory. Their work on the large flares of August 1972 remains a landmark achievement in the field.

After returning to Tokyo, he played a key role in the ASTRO-A project. The ASTRO-A satellite, which Tanaka named 'HINOTORI' (Japanese for Phoenix), observed more than 700 flares during its mission (February 1981 to October 1982). Combining the data from his X-ray spectrometer with the hard X-ray images taken by the X-ray telescope on HINOTORI, he classified X-ray flares into three types: (1) hot thermal flares, (2) impulsive flares, and (3) gradual hard flares. These differences could originate, he believed, from the environmental variety of flare sites. He summarized his view in an invited review paper in the Publications of the Astronomical Society of Japan in 1987. In 1985, Tanaka was awarded the Inoue Science Foundation Prize for his achievements in flare research.

Following the success of the HINOTORI project, Tanaka became a driving force in establishing the SOLAR-A (Yohkoh) project In 1983, however, he was diagnosed as suffering from fatal leukemia.

In 1988 he obtained a grant and started construction on a unique telescope to observe the magnetic field, velocity field, Hα, and white-light images of solar active regions simultaneously. He believed that the energy source of flares is the emergence of magnetic flux tubes, which are twisted (stressed) underneath the solar surface. His contribution to this issue of *Solar Physics* is the culmination of this research.

In Japanese, the name Katsuo means 'victory' and 'quickness', perhaps best symbolizing his great achievements in such a short life. We deeply regret the loss of this talented scientist.

E. HIEI and T. SAKURAI

Contents

THE YOHKOH (SOLAR-A) MISSION

(A Series of Invited Contributions)

THE SOLAR-A MISSION: AN OVERVIEW*

Y. OGAWARA, T. TAKANO, T. KATO

Institute of Space and Astronautical Science, Sagamihara, Kanagawa 229, Japan

T. KOSUGI, S. TSUNETA

Institute of Astronomy, University of Tokyo, Mitaka, Tokyo 181, Japan

T. WATANABE

National Astronomical Observatory, Mitaka, Tokyo 181, Japan

I. KONDO

Faculty of Systems Engineering, Shibaura Institute of Technology, Tameihara, Fukasaku, Omiya, Saitama 330, Japan

and

Y. UCHIDA

Department of Astronomy, Faculty of Science, University of Tokyo, Bunkyo-ku, Tokyo 113, Japan

(Received 10 April, 1991; in revised form 17 June, 1991)

Abstract. The SOLAR-A spacecraft is to be launched by the Institute of Space and Astronautical Science, Japan (ISAS) in August, 1991. As a successor of HINOTORI, this mission is dedicated principally to the study of solar flares, especially of high-energy phenomena observed in the X- and gamma-ray ranges. The SOLAR-A will be the unique space solar observatory during the current activity maximum period (1989–1992). With a coordinated set of instruments including hard X-ray and soft X-ray imaging telescopes as well as spectrometers with advanced capabilities, it will reveal many new aspects of flares and help better understand their physics, supporting international collaborations with ground-based observatories as well as theoretical investigations. An overview of this mission, including the satellite, its scientific instruments, and its operation, is given in this paper. Also the scientific objectives are briefly discussed.

1. Introduction

The SOLAR-A spacecraft is scheduled to be launched by the Institute of Space and Astronautical Science, Japan (ISAS) in August 1991 from the Kagoshima Space Center. The main goal of this spacecraft is to observe energetic phenomena related to solar flares in X- and gamma-rays with a coordinated set of instruments.

The first Japanese spacecraft dedicated to flare studies, HINOTORI, was launched close to the previous solar activity maximum in February 1981. At that time the Solar Maximum Mission (SMM) had been in operation for about a year, pursuing almost the same objectives as HINOTORI. The two missions successfully revealed new aspects of solar flares in X- and gamma-rays (e.g., Kundu and Woodgate, 1986; Tanaka, 1987).

The hard X-ray imagers on board the two spacecraft, for the first time, made images of flares in the hard X-ray range above a few keV. These novel observations showed

* After the launch the name of SOLAR-A has been changed to YOHKOH.

that in impulsive flares (type B of Tanaka, 1987) two or more separate sources brighten almost simultaneously in hard X-rays, and that extended hard X-ray sources (type C) are located high in the corona at altitudes of a few times 10^4 km above the photosphere in gradual X-ray flares.

The production and evolution of high-temperature plasmas in flares has been systematically studied via soft X-ray emission lines observed with high-resolution Bragg crystal spectrometers on board the P78–1, Tansei IV, SMM, and HINOTORI spacecraft. The precise diagnostics obtainable from various high-temperature emission lines revealed the dynamics of plasmas trapped in flaring loops; these include the violent heating of the transition region and the chromosphere due to precipitation of energetic electrons, turbulent motions, and evaporation (ablation) of the chromospheric material seen as blue-shifted components of emission lines in the initial phase of flares. A group of flares (type A) showed prominent hydrogen-like iron emissions from 'superhot' ($T > 3 \times 10^7$ K) thermal plasmas.

A number of gamma-ray-producing flares were also detected. It was found that ions can be accelerated to few-MeV energies within a few seconds and simultaneously with electrons at the flare onset. This has put strong constraints on theories of acceleration mechanisms. Gamma-ray events were preferentially observed near the solar limb, which suggests anisotropy of gamma-ray radiation and particle motions.

These discoveries from the last activity maximum, and their tantalizing physical interpretations, demanded further investigations. The images taken with the hard X-ray imagers had relatively low spatial resolution and their energy range was below about 30 keV so that the 'hard' X-ray images could be contaminated by X-rays from thermal sources. Another deficiency was the lack of reflecting telescopes for the soft X-ray wavelengths, where one can almost directly see the coronal magnetic structures in which flares occur. Such reflecting telescopes were flown previously only on SKYLAB in 1973–1974. The Bragg crystal spectrometers showed vividly the existence of chromospheric evaporation owing to high spectral resolution, but due to low sensitivities no clean data could be obtained at the very onset of a flare. Most gamma-ray flares were observed without simultaneous imaging at hard X-ray energies.

Based upon this scientific motivation, SOLAR-A was planned and constructed as an international collaborative project including many institutions in Japan, in the United States, and in the United Kingdom. The participating institutions and their responsibilities are summarized in Table I. In the present paper, we will give an overview of the SOLAR-A mission from the viewpoints of the scientific instruments in Section 2, the spacecraft design in Section 3, and the flight operations in Section 4. The scientific objectives will be briefly discussed in Section 5. *In this paper, as well as in the following series of papers in this Solar Physics issue, we will call this mission by its development name SOLAR-A, but it is to be noted that the mission will be formally given a new name after launch according to the ISAS tradition.*

TABLE I

The SOLAR-A mission

Mission objectives	Investigation of high-energy phenomena on the Sun
Launch	August 1991
Mission life	$\gtrsim 2$ years (orbital life $\gtrsim 3-4$ years)

Organization
    Project manager   Yoshiaki Ogawara (ISAS)

Project manager	Yoshiaki Ogawara (ISAS)
Project scientist	Yutaka Uchida (Univ. of Tokyo)
Principal investigators	
Hard X-ray Telescope (HXT)	Keizo Kai[a] (NAOJ) and Kazuo Makishima (Univ. of Tokyo)
Soft X-ray Telescope (SXT)	Tadashi Hirayama (NAOJ) and Loren W. Acton (LPARL; U.S. PI to NASA)
Wide Band Spectrometer (WBS)	Jun Nishimura (ISAS)
Bragg Crystal Spectrometer (BCS)	Eijiro Hiei (NAOJ) and J. Leonard Culhane (MSSL; U.K. PI to SERC)

Major participating institutions
  Japan: ISAS, National Astronomical Observatory (NAOJ), Univ. of Tokyo, Rikkyo Univ., Kyoto Univ., Nagoya Univ., etc.
  U.S. SXT team: Lockheed Palo Alto Research Laboratory (LPARL), Stanford Univ., U. California at Berkeley, and Univ. Hawaii
  U.K. and U.S. BCS team: Mullard Space Science Lab. (MSSL), Rutherford Appleton Lab., E.O. Hulburt Center for Space Research, and the National Institute of Standards and Technology

[a] Deceased 1991 March 11.

2. Scientific Instruments

SOLAR-A carries the following scientific instruments: the Hard X-ray Telescope (HXT), the Soft X-ray Telescope (SXT), the Wide Band Spectrometer (WBS), and the Bragg Crystal Spectrometer (BCS). As the technical details of these instruments are described in separate papers, only a brief summary is given below and in Table II.

2.1. HARD X-RAY TELESCOPE (HXT)

The HXT (Kosugi *et al.*, 1991) is a Fourier synthesis telescope. It consists of 64 bigrid modulation collimators, each with 2.3×2.3 cm cross-section and a 0.5 cm thick NaI(Tl) crystal attached to a 1-inch square phototube. The individual subcollimators measure spatially-modulated photon counts, thus providing 32 complex 'Fourier' components at appropriate position angles and wave numbers in the (u, v)-plane. The field of view covers the whole Sun, while the synthesis aperture, determined by the grid pitch corresponding to the fundamental wave number, is about 2×2 arc min. The angular resolution is about 5 arc sec. Images in four energy bands (15–24–35–57–100 keV) will be obtained simultaneously with a temporal resolution up to 0.5 s.

2.2. SOFT X-RAY TELESCOPE (SXT)

The SXT (Tsuneta *et al.*, 1991) is a grazing-incidence reflecting telescope in the soft X-ray band (3–60 A) with a CCD detector of 1024×1024 pixels. Its field of view covers

TABLE II

SOLAR-A scientific instruments

Hard X-ray Telescope (HXT)	
Instrument	Fourier-synthesis type collimator (64 elements)
Energy bands	15–24–35–57–100 keV (4 bands)
Angular resolution	~5 arc sec
Field of view	Full solar disk
Effective area	~70 cm^2
Time resolution	0.5 s
Soft X-ray Telescope (SXT)	
Instrument	Modified Wolter type I grazing incident mirror + CCD with coaligned optical telescope
Wavelength range (X-ray)	3–60 Å (selectable with filters)
(optical)	4600–4800 Å or 4293–4323 Å
Angular resolution	~2.5 arc sec
Field of view	Full solar disk
Time resolution	up to 0.5 s
Wide Band Spectrometer (WBS)	
Detectors	Gas proportional counter (soft X-rays; 2–30 keV)
	NaI scintillation counter (hard X-rays; 20–400 keV)
	BGO scintillation counter (gamma-rays; 0.2–100 MeV)
Time resolution	(count-rate data) 0.125, 0.25, or 0.5 s
	(pulse-height spectrum data) 1, 2, or 4 s
Bragg Crystal Spectrometer (BCS)	
Instruments	Bent crystal spectrometers
Spectral lines and resolutions	
S xv (5.0385 Å)	5.0160–5.1143 Å with 3.232 mÅ resolution
Ca xix (3.1769 Å)	3.1631–3.1912 Å with 0.918 mÅ resolution
Fe xxv (1.8509 Å)	1.8298–1.8942 Å with 0.710 mÅ resolution
Fe xxvi (1.7780 Å)	1.7636–1.8044 Å with 0.565 mÅ resolution
Time resolution	up to 0.125 s

the whole Sun. The angular resolution of the optical system of SXT is on the order of 2 arc sec across the solar disk, slightly better than that determined by the CCD pixel size (2.4 arc sec). Two filter wheels and a shutter device are placed in front of the CCD detector to choose energy band and exposure time properly. Filter and exposure selection as well as data acquisition are controlled by a dedicated microprocessor, following the commands dispatched by the main data processor on board the spacecraft. When the Sun is quiet, whole-Sun images are taken together with up to four bright active regions which are monitored at a moderate rate. When a flare occurs, the observation will be concentrated on the brightest region by taking 'partial-frame' images with up to 0.5 s time resolution.

2.3. WIDE BAND SPECTROMETER (WBS)

The WBS (Yoshimori *et al.*, 1991) consists of four types of detectors: the Soft X-ray Spectrometer (SXS), the Hard X-ray Spectrometer (HXS), the Gamma-Ray Spectrometer (GRS), and the Radiation Belt Monitor (RBM). The SXS, HXS, and GRS detectors observe the Sun, while RBM views perpendicular to the solar direction in order to monitor the radiation-belt environment of the spacecraft.

The SXS, a gas proportional counter filled with xenon, detects soft X-rays in the 2–30 keV band. Count-rate data are taken in two channels every 0.25 s and in a 128-channel pulse-height spectrum every 2 s. The HXS, a NaI(Tl) scintillation counter, detects 20–400 keV X-rays. Two-channel counting rates are taken every 0.125 s and 32-channel pulse-height spectrum every second. The GRS consists of two identical bismuth germanate (BGO) scintillators and detects 0.2–100 MeV gamma-rays, each producing six-channel count-rate data every 0.25 or 0.5 s and also a 128-channel pulse-height spectrum data every 4 s.

The RBM consists of a silicon diode detector and a NaI scintillator. The silicon detector measures counting rates of charged particles above about 20 keV every 0.25 s, while the NaI detector records counting rates in two channels every 0.25 s and 32-channel pulse-height data every second. In addition to solar flare observations, HXS and the NaI/bursts detector of RBM are used to detect cosmic gamma-ray burst during quiet periods of the Sun and during spacecraft night.

2.4. BRAGG CRYSTAL SPECTROMETER (BCS)

The BCS (Culhane *et al.*, 1991) consists of four bent-crystal spectrometers with position-sensitive proportional counters. The wavelength bands covering S xv, Ca xix, Fe xxv, and Fe xxvi lines are chosen to get information about the temperature and motion of hot plasmas produced in solar flares. Each energy band has up to 256 spectral bins. The BCS is equipped with its own queue memory (384 kbytes) to store the initial-phase data with high time resolution (up to 0.125 s). The data accumulation of BCS is controlled by a dedicated microprocessor to accommodate the maximum possible number of data with suitable time resolution during a flare. Data temporarily stored in the queue memory are read out at a fixed rate by the main data processor.

3. The Spacecraft

3.1. GENERAL

The SOLAR-A is to be launched in August 1991 from the Kagoshima Space Center at latitude 31 N, longitude 131 E, by an M-3S-II launcher, into a nearly circular orbit of about 600 km altitude, 31 deg inclination, and 97 min period.

The spacecraft, schematically shown in Figure 1, has dimensions of approximately $100 \times 100 \times 200$ cm with two external solar panels (150×200 cm each) outside, and weighs about 400 kg. The spacecraft body is made up of seven panels, i.e., one center panel and six surrounding panels. The center panel and two side panels form an H-shaped structure, the mechanical backbone of the spacecraft. The center panel holds the two large telescopes (SXT and HXT), as well as BCS, and plays the role of optical bench for them. The top panel facing the Sun holds the WBS detectors and also has

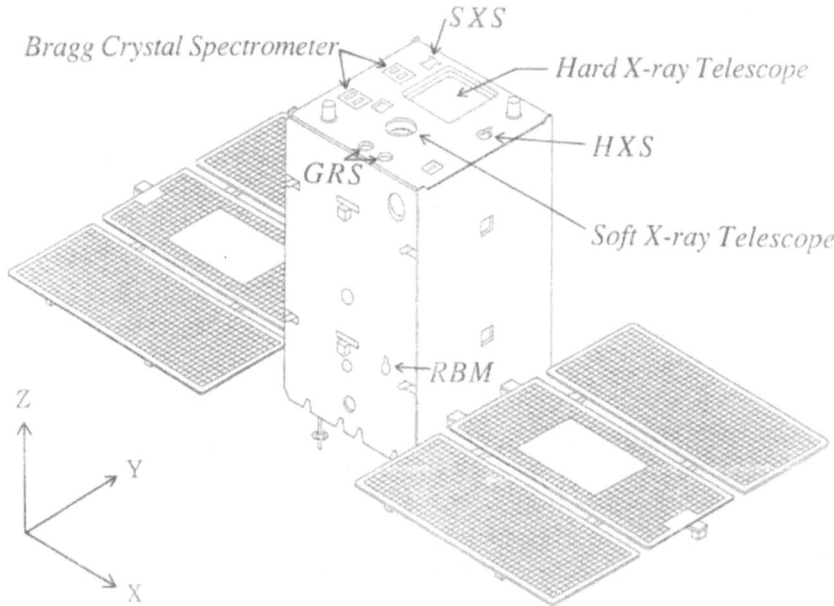

Fig. 1. The SOLAR-A spacecraft and its scientific instruments. The abbreviations used are: SXS, Soft X-ray Spectrometer; HXS, Hard X-ray Spectrometer; GRS, Gamma-Ray Spectrometer; and RBM, Radiation Belt Monitor. The spacecraft coordinates are given at the bottom left.

aperture windows for SXT, HXT, and BCS. Most of the electronics units are attached to the four side panels and the bottom panel.

Two solar cell panels are to supply about 570 W during spacecraft day. Excess power in daytime is stored into NiCd batteries for the power required during night. By this arrangement, about 220 W of power during day and 180 W during night are available.

The major parameters of the spacecraft are summarized in Table III.

3.2. ATTITUDE CONTROL

Since the SOLAR-A spacecraft is to take high spatial resolution images as well as spectra, precise control of the attitude is crucially essential. Thus it is stabilized in all three axes and its Z-axis (cf. Figure 1) is pointed at the center of the Sun with a stability of the order of 1 arc sec s^{-1} and several arc sec min^{-1}. The Y-axis is directed toward celestial north.

The attitude control system uses momentum wheels, magnetic torquers, and control-moment gyros as the actuators. As the attitude sensors, two Sun sensors and a star tracker, as well as geomagnetic sensors, are available for determining the spacecraft pointing relative to the direction of the Sun and to the ecliptic plane, respectively, while an inertial reference unit comprising four gyros detects changes of attitude with time. As a whole, the attitude system, using elaborate control programs with microprocessors, keeps the three axes well inside the requirements specified in Table III. The accuracy of Z-axis determination is estimated to be about 1 arc sec, disregarding bias error due

TABLE III

Major parameters of SOLAR-A

Size	~100 cm (L), 100 cm (W), 200 cm (H)
Weight	~400 kg
Power	~570 W (maximum; supply from solar cells)
Data recorder	10 Mbytes (magnetic bubble memory)
Data rate	32, 4, or 1 kbps
Telemetry rate	32, 4, or 1 kbps (real-time data)
	262 kbps (reproduced data from the data recorder)
Orbit	
Altitude	~600 km (nearly circular)
Inclination	~31 deg
Period	~97 min
Attitude control (requirement)	
Absolute pointing	a few arc min
Stability	around X/Y-axes (origin of image) <36 arc sec hr^{-1}
	<7 arc sec min^{-1}
	<1.2 arc sec s^{-1}
	around Z-axis (rotation of image) <5 arc min min^{-1}
	<20 arc sec s^{-1}
Z-axis pointing determination	$\lesssim 1$ arc sec
Offset pointing	capable, up to 45 arc min from the Sun center
Ground stations	
Commanding and downlink	Kagoshima Space Center (131 E, 31 N)
Downlink only	NASA stations at Goldstone, Madrid, and Canberra

to misalignment of the fine Sun sensor. Note that the two imaging telescopes have their own aspect sensors.

3.3. ONBOARD DATA PROCESSING

The scientific instruments of SOLAR-A, especially SXT, require sophisticated control of flight operations to exploit their capabilities within the constraints of the telemetry data rate and the capacity of the data recorder (cf. Table III). This is achieved by the 'data processor' (DP) unit.

The DP consists of dual redundant microcomputer systems. Further, the most essential parts of the DP functionality can also be achieved by hardwired logic if both of the microcomputer systems should fail to operate.

The fundamental functions of DP are as follows:

(1) Data gathering from all the instruments.

(2) Data processing and editing into the telemetry stream.

(3) Data recording to and data dump from the data recorder, including complicated control of the data recorder (cf. Section 4.3).

(4) Automated control of observing mode, the data rate, and the operation of the scientific instruments, especially SXT. The mode and rate are switched by DP depending on conditions such as occurrence of flares, spacecraft sunrise/sunset, etc.

The data can be telemetered to the ground in real time, with data rates of 32, 4, or 1 kbps, depending on the observation conditions. Since contacts with downlink stations are limited, the data are also stored in an on-board recorder, and dumped during ground-station contacts. The recorder is a magnetic bubble data recorder with 10-Mbyte capacity. Unfortunately this capacity is not sufficient for recording data continuously at the high rate (32 kbps) for a full orbital period, so that sophisticated control algorithms are required. Such operations-related items, together with data processing and editing, will be further discussed in Section 4.

3.4. COMMAND SYSTEM

An uplink commanding system controls the operation of all the instruments on the spacecraft. Commands from the ground are to be sent only from the Kagoshima Space Center, during contacts of about 10 min duration each for 5 orbits day^{-1}. Each command is distributed by the telemetry command control unit to the instrument specified by the instrument code included in the command.

In addition, the control unit can coordinate sequences of commands. First, it can store up to 128 sets of 'organized commands' (OG's), each being a set of up to 32 commands. An OG can be launched by an 'OG start command' from the ground. Also several OG's can be automatically triggered to start at the times of interruption messages issued by specific on-board instruments. Such interruptions include sunrise/sunset (from a Sun sensor), occurrence of a flare (from DP), and emergencies in the power control system or the attitude control system. Second, a series of OG's can be dispatched sequentially with specified time intervals. Such a command series is called an 'operation program (OP)', and is controlled by a program which contains a sequence of 128 OG addresses and intervals stored in the program memory. The operation program is initiated by an 'OP start command' and can last for up to about 10 days, so that the operator can easily program several days' spacecraft operation beforehand.

3.5. TELEMETRY

Data acquired with the instruments on board SOLAR-A can be telemetered to three Deep Space Network stations at Goldstone, Madrid, and Canberra, as well as to the Kagoshima Space Center station. Two telemetry channels are used, one at S-band (2.2 GHz) and the other at X-band (8.4 GHz). At Kagoshima, the two channels are received simultaneously; the S-band channel transmits 'real-time data', while the X-band transmits 'reproduced data' from the bubble data recorder at 262 kbps. The 10-Mbyte data stored in the data recorder can be sent in about 5 min well within the 10-min contact duration. On the other hand, at the other stations only reproduced data are transmitted via the S-band and no real-time data are available.

Real-time data downlinked at Kagoshima are sent to ISAS at Sagamihara, near Tokyo, via a real-time data link. Reproduced data are also sent to ISAS within 30 min from downlink. Data taken at the Deep Space Network stations are brought first to the Jet Propulsion Laboratory and then transferred to ISAS through a NASA Communication (NASCOM) line in less than a few days. For archiving and scientific analysis all of the data are finally stored and maintained in a database system at ISAS. For more information the reader should refer to the article by Morrison *et al.* (1991).

4. Operations

Since the primary objective of SOLAR-A is to understand the high-energy aspects of solar flares, it is of crucial importance to observe as many flares as possible. Major flares are given higher priority, because such flares emit a wide range of radiations from soft X-rays to gamma-ray lines and thus make comprehensive studies possible. At the same time, the mission objectives include understanding of preflare conditions of active regions and also non-flare phenomena such as coronal holes, interconnecting loops, X-ray bright points, etc. The SOLAR-A instruments, especially the Soft X-ray Telescope (SXT), have good capabilities for making such preflare and non-flare observations. In the following we will discuss how the SOLAR-A program has been organized to accomplish these two different (and sometimes mutually contradictory) types of observations in an automated but flexible manner.

4.1. OBSERVING MODES

The spacecraft has four observing modes ('flare', 'quiet', 'night', and 'BCS-out' modes) and three telemetry data rates ('high' 32 kbps, 'medium' 4 kbps, and 'low' 1 kbps). The flare and quiet modes run either at high or medium bit rate. The night mode, used in spacecraft night, runs at low rate. The BCS-out mode at high rate, used specifically for sweeping out the BCS queue memory, is initiated when the spacecraft enters night or just after the flare mode ceases.

Each mode has its own telemetry data format, i.e., the telemetry assignment differs from one mode to another, as shown in Table IV. There is only one significant difference between the quiet mode and the flare mode; the four scientific instruments (HXT, SXT, WBS, and BCS) share the telemetry in flare mode, while HXT yields most of its telemetry assignment to SXT in quiet mode.

TABLE IV
Mode and telemetry assignment in each 128-byte frame

	Flare mode 32 or 4 kbps	Quiet mode 32 or 4 kbps	BCS-out mode 32 kbps	Night mode 1 kbps
Basic data[a]	32 bytes	32 bytes	32 bytes	32 bytes[b]
HXT	16	–	–	–
SXT	64	64 + 16[c]	–	–
WBS	8	8	–	–
BCS	8	8	64	–

[a] Contains the spacecraft system information, status and housekeeping data, and some basic monitoring data from the scientific instruments.
[b] Remaining 96 bytes are used for recording cosmic gamma-ray bursts.
[c] Two sections.

In quiet mode the two SXT telemetry sections (64 + 16 bytes) are assigned to obtain 'full frame' images (whole-Sun images) together with what we call 'partial frame' images (images with areas small compared to the whole CCD field of view) in parallel, i.e., one of the two is used for whole-Sun images and the other for partial-frame images. If the 16-byte section is used for whole-Sun images, the image cadence is typically once every 10 min at high data rate (assuming 2×2 on-chip summation mode; see Tsuneta *et al.* (1991) for more details). This time resolution can be improved by a factor of 4 by assigning the 64-byte section to whole-Sun images, resulting in a reduced area size and/or reduced time resolution for partial-frame images. In flare mode, on the other hand, only partial-frame images are telemetered, with no whole-Sun images. The time resolution is 2 s if the area size is 64×64 pixels (about 2×2 arc min) at high data rate.

The 'basic' section of the telemetry data format is not affected by the observing mode. This contains the spacecraft system information (clocks, command answerback, observing mode/data rate, attitude, power supplies, etc.), status and housekeeping data for all of the instruments, and also some basic monitoring data from the scientific instruments. With regard to HXT, data from the lowest-energy band (15–24 keV) are always recorded in this section, so that we can expect to obtain preflare hard X-ray images even before the flare mode is initiated by a flare. Note that at higher energies usually no hard X-rays with intensity high enough to be image-synthesized are emitted without the occurrence of a flare.

The observing modes also affect the control of the instruments. For example, the SXT control in flare mode is completely independent of that in quiet mode (for details see the separate papers in this issue).

4.2. AUTOMATED MODE CONTROL

The logic of the DP software that controls the observing modes is schematically shown in Figures 2 and 3.

At the beginning of each spacecraft day the operating mode is 'quiet' and the data rate is predetermined at 'high' or 'medium'. In this mode SXT takes whole-Sun images to monitor the global structure of the corona together with partial-frame images of active regions to monitor their development. The data taken with HXT are restricted to the lowest energy band.

When DP recognizes flare occurrence by an abrupt increase of counting rate in one of the three sensors (HXS, SXS, and BCS; selectable) above a 'flare threshold', it turns on the flare flag and initiates flare mode at high rate within the next two seconds, provided that there is no simultaneous increase in the RBM counting rates which monitor the particle background environment. In this mode the whole HXT data are sent to the telemetry, while for SXT image-processing software picks out the brightest region and edits partial-frame images for telemetry.

This mode continues unconditionally during a preset 'flare minimum duration'. After this period elapses, two thresholds are applied once every 64 s to determine the mode and data rate. If the counting rates exceed the 'great flare threshold' (case C of Figure 3), the flare mode continues at high rate. If the flare flag turns off due to lower counting

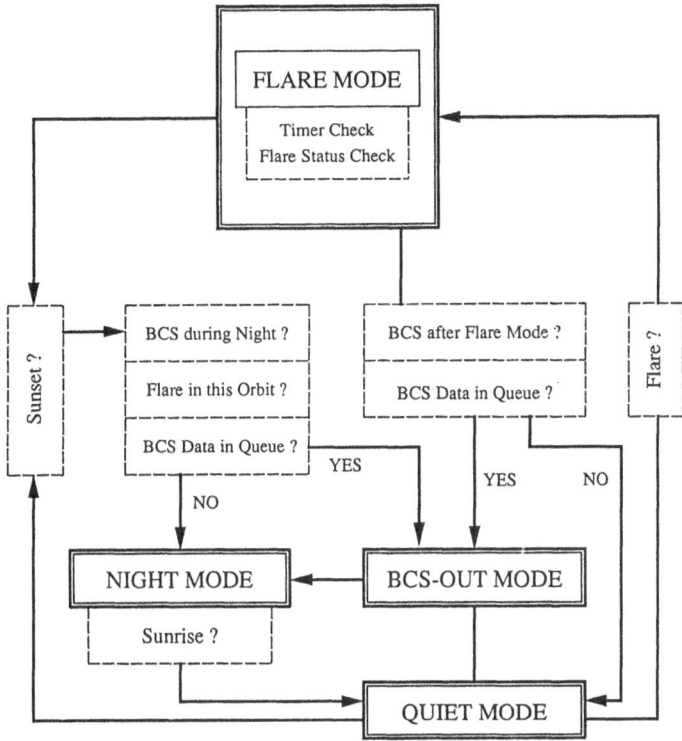

Fig. 2. Diagram showing the logic of automated observing mode control of SOLAR-A. The on-board microcomputer (DP) continuously monitors the day/night signal provided from a Sun sensor, various flare sensor (HXS, SXS, or BCS) counting rates for setting/resetting the flare flag, and the BCS queue memory status. The RBM counting rates are also monitored (not shown) for judging whether an increase of flare sensor counting rate is due to a solar flare or to trapped radiation.

rates than the 'flare end threshold' (case A of Figure 3), the mode is changed to quiet. Otherwise (case B of Figure 3), the flare mode continues but the data rate is changed to medium. This allows recording any long-enduring flare until its end.

When flare mode ends, the observing mode usually returns to quiet. However, if so specified in advance, data stored in the BCS queue memory are swept out (BCS-out mode) before the quiet mode is initiated. When the spacecraft enters into the shadow of the Earth, the mode is changed into night mode, in which data taken with HXS and RBM are recorded to monitor occurrence of cosmic gamma-ray bursts and the particle environment of the spacecraft. The BCS-out mode can be initiated before night mode if its queue memory stores flare data.

Besides this automated operation, it is possible to control the observing mode manually either by directly specifying the mode or by indirectly setting/resetting flags. Such a manual control is initiated and terminated either by commands from the ground (when the spacecraft is within the reach of the Kagoshima Space Center) or by the operation program (see Section 3.4). This manual mode control function will be used, for example, for calibrating the instruments.

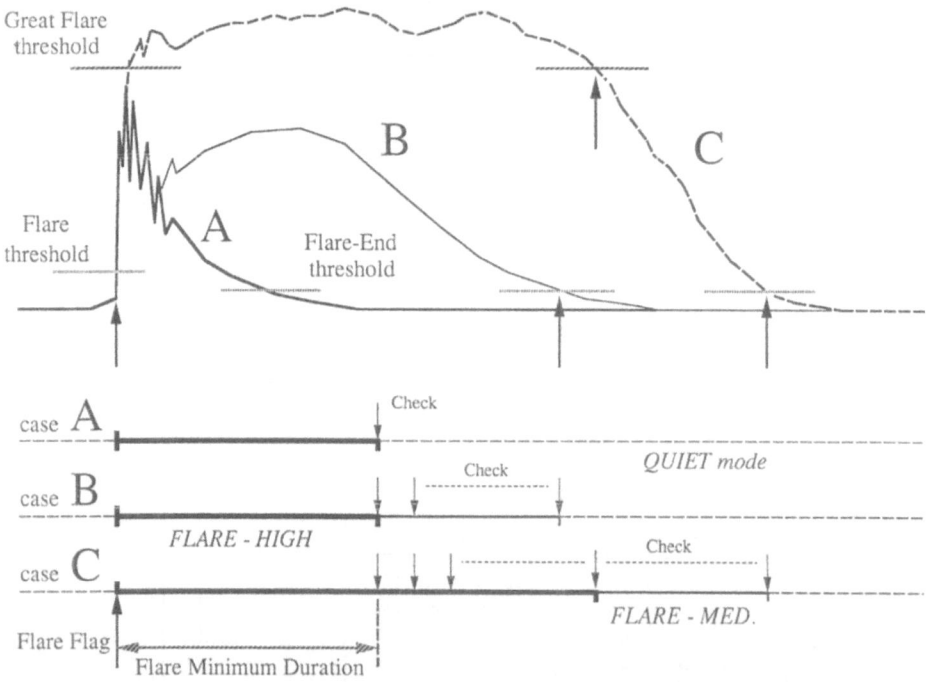

Fig. 3.   Three flare thresholds and the control of the flare mode. The flare and flare-end thresholds are used for setting and resetting the flare flag, respectively. The flare mode starts at high data rate. After a specified time elapses, the mode and data rate depend on the flare sensor counting rates. The 'great flare threshold' allows classification of flares into two classes, and determines the data priority against overwrite on the data recorder (see text).

4.3. CONTROL OF DATA STORAGE TO THE BUBBLE DATA RECORDER

Because flare occurrence cannot be scheduled, an on-board data recorder is essential. However, the data can consume the bubble data recorder (BDR) capacity of 10 Mbytes in some 40 min at high rate, so that overwriting is inevitable. This causes a serious problem because one orbit day lasts for more than 60 min. This problem would become more serious in cases when some of the downlink stations were not available.

To avoid loss of important data by overwriting with less important data, the DP software is so designed as to compare the importance between the data which DP is going to write and the data which is already stored, and, if necessary, to protect the stored data by skipping.

The priority levels usually used are as follows:

'great flare' data > 'normal flare' data > quiet or night mode data .

Here 'great flares' are flares whose peak counting rates exceed the 'great flare threshold'. This is the same logic as adopted in the HINOTORI mission which resulted in successful observations of many major flares (Kondo, 1983). It is to be noted, however, that in the case of SOLAR-A this overwrite control is executed by the DP software so that it is more flexible and more powerful. The logic can protect not only flare data but also any associated preflare data. The priority levels are changeable. Further, the observer can unconditionally protect the data taken in a certain period of time by declaring a 'campaign observation'. This campaign observation function will be used to protect, for example, SXT whole-Sun images taken at a regular interval, say once in a few orbits, used for example for a 'synoptic movie' in a given filter. This function will be also useful when the SOLAR-A team makes a short-term, simultaneous observation with certain observatories in which the loss of data by overwriting would be undesirable.

5. Scientific Topics

The SOLAR-A mission is a unique space solar observatory accommodating a coordinated set of instruments. The instruments are individually advanced over their predecessors, but there is no doubt that their combination together will make scientific return more fruitful.

For example, the HXT imaging of hard X-ray sources with high time resolution in the four energy bands up to 100 keV will give us information about rapid processes occurring in the impulsive phase of a flare, as well as fainter thin-target sources showing the location of the storage of accelerated particles. Coaligned flare images taken simultaneously with SXT will be of crucial importance to locate the hard X-ray sources on the flaring loop structure seen in soft X-ray images. High-resolution images in the two energy ranges will give much-improved physical interpretations of the energy release and particle acceleration processes in solar flares. In this respect, we note that the two imagers have wide fields of view that cover the whole Sun, which guarantees simultaneous observations for most flares. Similarly the use of BCS together with SXT will allow us to discern the successive brightenings of neighboring points from the true motion of the X-ray emitting plasma.

Another advantage of SOLAR-A is that fainter objects such as loop structures in the quiet corona are expected to be observable with SXT due to its high sensitivity and wide dynamic range. The preflare evolution of active regions is an interesting objective to be examined in detail. Such non-flare objectives may include globally-distributed X-ray bright points, possibly related to the internal dynamo activity of the Sun. Another example would be the morphology of faint coronal loops that we expect to reveal the magnetic connectivity of the outer atmosphere of the Sun and the solar wind.

In the following, we briefly mention possible scientific objectives of SOLAR-A, and give comments from these viewpoints. Since a thorough survey and discussion of the objectives of SOLAR-A are beyond the scope of this introductory overview, we simply make a candidate list and give a very brief discussion of a single topic as an example.

A list of objectives may include:

(A) *Flare-related phenomena*

(1) Evolution of active regions, especially their preflare states, in terms of the coronal loop structures that show their magnetic connectivity and its changes.

(2) Flare onset: what happens immediately before and at the very beginning of the impulsive phase?

(3) High-temperature loops and arcades: when and how are they formed and where do the mass and energy come from?

(4) Electron accelerations: where and when exactly?

(5) Ion accelerations and gamma-ray line flares: where and when?

(6) Dynamic behavior of the footpoints of the flaring loops: mass ejection and/or evaporation from the chromosphere.

(7) Flare ejecta, shocks, and plasmoids: are they observable in X-rays?

(8) Relation between the sources of hard X-ray, soft X-ray, optical, and radio emissions.

(9) White-light flares (observable with the SXT aspect sensor).

(B) *Dynamical phenomena not necessarily related to flares*

(1) Surges and Brueckner's jets in X-rays.

(2) Disappearances of quiescent filaments and related phenomena (in some cases, accompanying low-energy two-ribbon flares).

(3) Coronal mass ejections and related phenomena.

(C) *Other forms of activity*

(1) X-ray bright points and their solar-cycle variations.

(2) Micro- and nano-flares.

(3) Formation and evolution of active-region loops.

(D) *Global coronal structure, and others*

(1) Formation and evolution of quiet coronal loops, if possible.

(2) Behavior of coronal holes.

(3) Solar oscillations (observable with the SXT aspect sensor).

(4) Other kinds of simple photospheric imaging.

Here we pick only one example from this list, namely, the problem of what is happening immediately before the 'flare onset'.

HINOTORI and SMM indicated (a) that evidence exists for the presence of a very important stage, in which violent uprising motions with 300–400 km s^{-1} were found in Fe XXV or Ca XIX lines immediately before the flare onset (Tanaka *et al.*, 1983; Antonucci, 1983); (b) that an X-ray source brightens up and is confined at the top of a loop-like structure, already with high temperature when it first appears in X-rays (Tsuneta *et al.*, 1984), without rapid expansion or rapid cooling, requiring special mechanisms of confinement and insulation, and (c) that gamma-ray line emissions appear simultaneously with hard X-ray impulsive bursts (Forrest and Chupp, 1983; Nakajima *et al.*, 1983), requiring almost instantaneous ion acceleration if the impulsive phase is taken to indicate the flare onset. Observations (a) and (b) suggest the possibility that a very important dynamical phase, occurring immediately before the impulsive phase and possibly carrying mass and energy from below, has escaped our attention thus far (Uchida and Shibata, 1988). If this turns out to be the case by observations with SOLAR-A, it would change our understanding of flares drastically. Note that, in this case, even observation (c) does not necessarily mean any instantaneous acceleration of ions. Instead it would mean the simultaneous release of both electrons and ions accelerated in a dynamical phase before the impulsive phase. The clarification of the processes occurring in this period would be vital for our understanding of flares.

This example shows that an obvious contribution which SOLAR-A can make will be to give a precise answer to the questions regarding where, when, and how mass and energy actually enters a flaring coronal loop. Whether magnetic reconnection really occurs or not can also be examined by SOLAR-A because it can give us the magnetic connectivities and their changes in the flaring locations in terms of coronal loop structures in the preflare and flaring states.

6. Concluding Remarks

There will be no other satellite project totally dedicated to solar flare observations during the current maximum period. Only the SOLAR-A mission will provide systematic data covering the broad energy range from soft X-rays to gamma-rays. The coordinated set of instruments on board this mission will provide precious and well-organized data to investigate solar flares and related high-energy phenomena in the corona. The instruments make substantial improvements over those flown earlier on spacecraft such as HINOTORI and SMM.

Although observations in X- and gamma-rays are well established to be very important in the study of solar flares, they still provide only partial information about the flare phenomenon. Solar flares have many aspects, and collaborative observations with ground-based optical and radio telescopes will be extremely important in clarifying the physics of flares and of other active phenomena. In addition, theoretical investigations that can help to synthesize a coherent picture from these complex data will be indispensable. The SOLAR-A will take part in these coordinations and contribute to the investigation of such thorny problems.

We believe that the fruitful scientific return from SOLAR-A will unfold still newer aspects of our understanding of flares and that these new aspects will lead us to develop a more advanced space observatory which, we hope, will be flown as SOLAR-B at some time in the future.

Acknowledgements

The authors wish to acknowledge the staff of ISAS, and engineers of many companies involved in this project. All the team members express their thanks to the National Aeronautics and Space Agency (NASA) and to the Science and Engineering Research Council (SERC) for their support and contribution to this project. Special thanks are due to Professors Yasuo Tanaka and Minoru Oda, without whose support the

SOLAR-A mission would not have existed. Two important persons passed away during the preparatory stage of this mission. We express our thanks to the late Prof. Katsuo Tanaka, who first proposed the SOLAR-A mission as a successor of HINOTORI, and the late Prof. Keizo Kai, who worked hard until the end as Principal Investigator for HXT. Hugh Hudson and Bob Bentley are acknowledged for their assistance in preparing this manuscript.

References

Antonucci, E.: 1983, *Solar Phys.* **86**, 67.

Culhane, J. L., Hiei, E., Doschek, G., Cruise, A. M., Ogawara, Y., Uchida, Y., Bentley, R. D., Brown, C. M., Lang, J., Watanabe, T., and 17 co-authors: 1991, *Solar Phys.* **136**, 89 (this issue).

Forrest, D. J. and Chupp, E. L.: 1983, *Nature* **305**, 291.

Kondo, I.: 1983, in Y. Tanaka *et al.* (eds.), *Proceedings of Hinotori Symposium on Solar Flares*, ISAS, Tokyo, p. 3.

Kosugi, T., Makishima, K., Murakami, T., Sakao, T., Dotani, T., Inda, M., Kai, K., Masuda, S., Nakajima, H., Ogawara, Y., Sawa, M., and Shibasaki, K.: 1991, *Solar Phys.* **136**, 17 (this issue).

Kundu, M. R. and Woodgate, B. (eds.): 1986, *Energetic Phenomena on the Sun*, The Solar Maximum Mission Flare Workshop Proceedings, NASA-CP 2439.

Morrison, M., Lemen, J. R., Acton, L. W., Bentley, R. D., Kosugi, T., Tsuneta, S., Ogawara, Y., and Watanabe, T.: 1991, *Solar Phys.* **136**, 105 (this issue).

Nakajima, H., Kosugi, T., Kai, K., and Enome, S.: 1983, *Nature* **305**, 292.

Tanaka, K.: 1987, *Publ. Astron. Soc. Japan* **39**, 1.

Tanaka, K., Nitta, N., Akita, K., and Watanabe, T.: 1983, *Solar Phys.* **86**, 91.

Tsuneta, S., Nitta, N., Ohki, K., Takakura, T., Tanaka, K., Makishima, K., Murakami, T., Oda, M., and Ogawara, Y.: 1984, *Astrophys. J.* **284**, 827.

Tsuneta, S., Acton, L., Bruner, M., Lemen, J., Brown, W., Caravalho, R., Catura, R., Freeland, S., Jurcevich, B., Morrison, M., Ogawara, Y., Hirayama, T., and Owens, J.: 1991, *Solar Phys.* **136**, 37 (this issue).

Uchida, Y. and Shibata, K.: 1988, *Solar Phys.* **116**, 291.

Yoshimori, M., Okudaira, K., Hirasima, Y., Igarashi, T., Akasaka, M., Takai, Y., Morimoto, K., Watanabe, T., Ohki, K., Nishimura, J., Yamagami, T., Ogawara, Y., and Kondo, I.: 1991, *Solar Phys.* **136**, 69 (this issue).

THE HARD X-RAY TELESCOPE (HXT) FOR THE
SOLAR-A MISSION*

T. KOSUGI[1], K. MAKISHIMA[2], T. MURAKAMI[3], T. SAKAO[4], T. DOTANI[3],
M. INDA[2], K. KAI[4], S. MASUDA[1], H. NAKAJIMA[4], Y. OGAWARA[3],
M. SAWA[4], and K. SHIBASAKI[4]

(Received 22 March, 1991; in revised form 27 May, 1991)

Abstract. The Hard X-ray Telescope (HXT) is a Fourier-synthesis imager; a set of spatially-modulated photon count data are taken from 64 independent subcollimators and are Fourier-transformed into an image by using procedures such as the maximum entropy method (MEM) or CLEAN. The HXT takes images of solar flares simultaneously in four energy bands, nominally 15 (or 19)–24, 24–35, 35–57, and 57–100 keV, with an ultimate angular resolution as fine as ~ 5 arc sec and a time resolution 0.5 s. Each subcollimator has a field of view wider than the solar disk. The total effective area of the collimator/detector system reaches ~ 70 cm², about one order of magnitude larger than that of the HINOTORI hard X-ray imager. Thanks to these improvements, HXT will for the first time enable us to take images of flares at photon energies above ~ 30 keV. These higher-energy images will be compared with lower-energy ones, giving clues to the understanding of nonthermal processes in solar flares, i.e., the acceleration and confinement of energetic electrons. It is of particular importance to specify the acceleration site with regard to the magnetic field figuration in a flaring region, which will be achieved by collaborative observations between HXT and the Soft X-ray Telescope on board the same mission.

1. Introduction

The SOLAR-A mission is the second satellite of the Institute of Space and Astronautical Science, Japan (ISAS) dedicated to the study of solar flares. The first solar-flare mission, HINOTORI, was launched during the previous maximum in February 1981. At that time the Solar Maximum Mission (SMM) had been in operation for about one year with almost the same scientific objectives as HINOTORI. Each of the two missions conducted a wide variety of observations in the X- and γ-ray ranges (including EUV and visible-light observations in the case of SMM), and diagnosed high-energy particles and high-temperature plasmas involved in flares. In particular, the two missions made the first hard X-ray imaging observations of flares (Van Beek *et al.*, 1980; Makishima, 1982).

Hard X-rays are produced by collisions of high-energy electrons with ions (bremsstrahlung) and their propagation is almost unaffected by the solar atmosphere above the photosphere. Accordingly, images taken in this range provide direct information on the generation, transfer, and confinement of high-energy electrons. In fact, the two earlier experiments revealed that hard X-ray imaging observations were very promis-

[1] Institute of Astronomy, University of Tokyo, Mitaka, Tokyo 181, Japan.
[2] Department of Physics, Faculty of Science, University of Tokyo, Bunkyo-ku, Tokyo 113, Japan.
[3] Institute of Space and Astronautical Science, Sagamihara, Kanagawa 229, Japan.
[4] National Astronomical Observatory, Mitaka, Tokyo 181, Japan.

* After the launch the name of SOLAR-A has been changed to YOHKOH.

ing, as expected. At the same time, the experiments should be regarded as preliminary ones; the two imagers had only one or two energy bands in the hard X-ray range, all below ∼ 30 keV, and the angular and temporal resolutions were not sufficient. Thus, it is evident that an advanced hard X-ray imager will play a very important role to settle the puzzles of solar flares (e.g., Dennis, 1988). The Hard X-ray Telescope (HXT) for the SOLAR-A mission is such an instrument.

In the following, we will give an overview of HXT and its scientific objectives in Section 2. These objectives and the expected properties of the sources lead to X-ray optical design and the image synthesis principle (Section 3). Section 4 gives an outline of the HXT instrument. A few remarks will be made in Section 5 with regard to the performance of the flight instrument.

2. HXT Overview

The HXT is a Fourier-synthesis imager of 64 elements; each subcollimator measures a spatially-modulated incident photon count. A set of photon count data from the 64 subcollimators are telemetered to the ground and synthesized there into an image through, at least in principle, the Fourier transform.

Instrumentally HXT consists of three major sections (Figure 1).

The collimator (HXT-C) is the X-ray optics part of the instrument. Simply it is a metering tube (417 mm × 376 mm × 1400 mm) with X-ray grid plates at both ends. Each grid plate is an assembly of 64 subcollimator grids made of tungsten 0.5 mm thick. At the center of the X-ray optics is installed the aspect system (HXA) optics, which include lenses with appropriate filters on the front grid plate and fiducial marks on the rear plate, thus providing white-light images of the Sun which yield the axis direction information of the X-ray optics with respect to the solar disk. The HXT-C weighs ∼ 13.5 kg.

The detector assembly (HXT-S; S denotes sensors) is a package of 64 detector modules (465 mm × 392 mm × 223 mm, weight ∼ 17.1 kg); each module consists of a NaI(Tl) scintillation crystal (25 mm square) and a photomultiplier tube, with dynode bleeder string and pre-amplifier assembled as a unit. Eight high-voltage power supply units are attached to HXT-S. The HXT-S also has two one-dimensional CCD's at the center for detecting HXA visible-light signals.

The electronics unit (HXT-E) is 374 mm × 246 mm × 220 mm in size and ∼ 10.8 kg in weight. It processes hard X-ray signals from HXT-S; first it converts pulse-height analogue signals into digital signals and then it counts the incident photon number after discriminating the photon energy into four energy bands. Here, signals from the individual subcollimators are processed separately and simultaneously. Finally HXT-E sends the photon count data to the data processor (DP; Ogawara *et al*, 1991) of the spacecraft once every 0.5 s. The HXA signals are also digitized and processed in HXT-E. In addition HXT-E plays the role of power/mode controller of the whole HXT instrument.

The main characteristics of HXT are summarized and compared with its two pre-

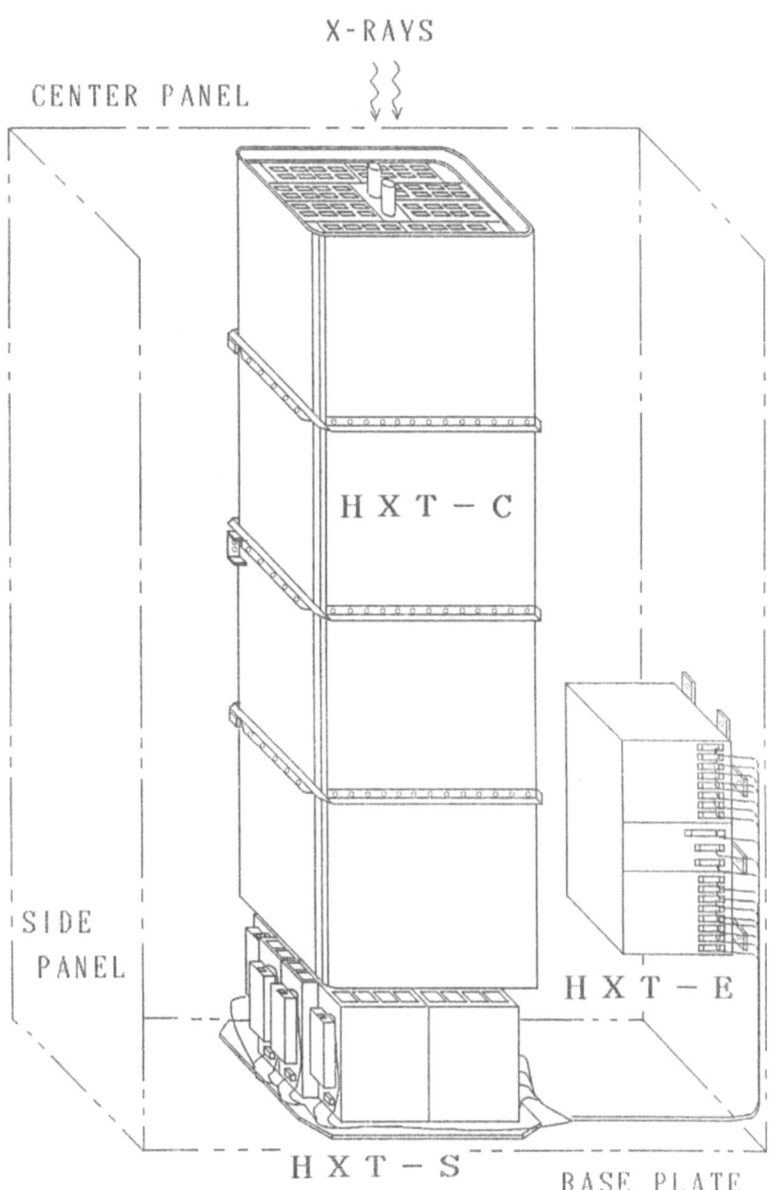

SOLAR — A HXT

Fig. 1. Schematic drawing of the HXT instrument. HXT consists of three major sections: the collimator (HXT-C), the detector assembly (HXT-S), and the electronics unit (HXT-E). The aspect system (HXA) is installed along the central axis of HXT-C and HXT-S.

TABLE I

Main characteristics of HXT (comparison with its predecessors)

	HXT (SOLAR-A)	HINOTORI imager	HXIS (SMM)
Collimator type	Multi-el. bigrid MC	Rotating bigrid MC	Multi-el. IC
No. of elements	64 SC's	2 (orthogonal)	(F)304; (C)128
Size of elements	23 mm ☐	120 mm Ø	7.5 mm Ø
Image acquisition	2D Fourier synthesis	1D scans → 2D image	1 el./1 pixel
Angular resolution	~5″	~10″	8″ (32″)
Field of view	whole Sun	whole Sun	2′40″ (6′24″)
Synthesis aperture	2′06″	2′12″	2′40″ (6′24″)
Time resolution	0.5 s	~10 s	1.5–9 s
Energy bands	4 channels	1 channel	6 channels
(keV)	Ch. L: 15(19)–24		3.5–5.5–8.0
	Ch. M1: 24–35	5(17)–40	8.0–11.5–16
	Ch. M2: 35–57		16–22–30
	Ch. H: 57–100		
Detector	NaI(Tl) scint.	NaI(Tl) scint.	Gas prop. counter
	(25 mm ☐ × 64)	(120 mm Ø × 2)	(7.5 mm Ø × 900)
Effective area	~70 cm^2	~8 cm^2 × 2	0.07 cm^2 pixel^{-1}

Note: MC = modulation collimator; IC = imaging collimator; SC = subcollimator; 2D = two-dimensional; 1D = one-dimensional; (F) = fine field of view; (C) = coarse field of view.

decessors in Table I. As can be seen from the table, HXT has several advantages over its predecessors. These include:

– higher energy range to ~100 keV,
– higher sensitivity,
– improved angular and temporal resolutions,
– a wide field of view covering the whole Sun, and
– simultaneous observation with the Soft X-ray Telescope (SXT; Tsuneta et al., 1991) on board the same mission.

With these advanced characteristics, HXT will make the first imaging observations in the ≳ 30 keV hard X-ray range, collect many examples of hard X-ray flares, and help to answer the following specific questions about solar flares:

(1) Where are electrons accelerated in flaring magnetic flux loops: inside a highly-stressed single loop or at the interaction site of multiple loops?

(2) How does the acceleration process take place? Are electrons accelerated in a direction parallel to the magnetic lines of force or perpendicular? Or are electrons energized randomly? In what conditions does the process work?

(3) What causes the double-source structure of hard X-ray sources in impulsive (type B) flares? Do down-streaming electron beams exist and play an important role?

(4) Is there a different acceleration process operating in gradual hard (type C) flares from that of impulsive (type B) flares? What makes hard X-ray sources appear high in the corona?

(5) What determines the type of a flare? Is it a plasma parameter such as density or plasma β, or the magnetic field topology?

(6) Are electrons and ions accelerated simultaneously by a single process or separately? Are ions accelerated in a specific type of flare? If ions are accelerated in a different phase or step from electrons, what causes the second phase or step? (This topic will be pursued with simultaneous observations with the Gamma-Ray Spectrometer (GRS; Yoshimori et al., 1991) on board the same mission.)

(7) How are the energized electrons confined in a magnetic flux loop or an arcade of loops? Does reacceleration of electrons take place?

(8) As a whole, what role does the acceleration of particles play in a flare? Is it a minor episode in a complicated flare process, or is it the main cause of subsequent thermal responses of the flaring solar atmosphere?

3. Design and Image Synthesis Principles

The HXT instrument is the world's first X-ray telescope that adopts the Fourier synthesis principle for taking images. This type of telescope was first discussed by Makishima et al. (1978) as multi-pitch modulation collimator (MPMC). Recently, in the course of design study for an advanced flare-oriented hard X-ray telescope, Prince et al. (1988) proposed a Fourier-transform telescope using position-sensitive detectors, each using one subcollimator and measuring one complex Fourier component.

The HXT is a Fourier-synthesis telescope with a novel design principle. Each subcollimator has a normal, non-position-sensitive detector and measures only one of a Fourier component pair. Here we use the term Fourier component for simplicity. Note that here and in the following the term does not have the usual sense – a coefficient in a series of trigonometric expansion of a function – but is generalized.

The design principle of HXT is based upon the fact that a single modulation collimator, with two identical grids having pitch (slit spacing) twice the slit width, gives a transmission function that is a repetitive triangular pattern. The function resembles the trigonometric (cosine or sine) function if we disregard the DC component involved. A brief explanation of the principle is given in Figure 2 and in what follows.

Let us represent the transmission function of one of a pair of subcollimators by

$$F_c(k\rho) ; \qquad \rho = X \cos \theta + Y \sin \theta,$$

where k ($= 1, 2, \ldots$) denotes the wave number, $k\rho$ the modulation phase, θ the position angle of the grid pattern, and X and Y the spatial coordinates normalized to the fundamental ($k = 1$) period of repetition. In addition to this first subcollimator of the pair, there is a second whose position angle and pitch are the same as those of the first but whose relative slit positions are shifted by a quarter of the pitch. The transmission function of the second is then

$$F_s(k\rho) = F_c(k\rho - \pi/2),$$

i.e., the pattern is the same as the first, but its phase is shifted by $90°$ from the first.

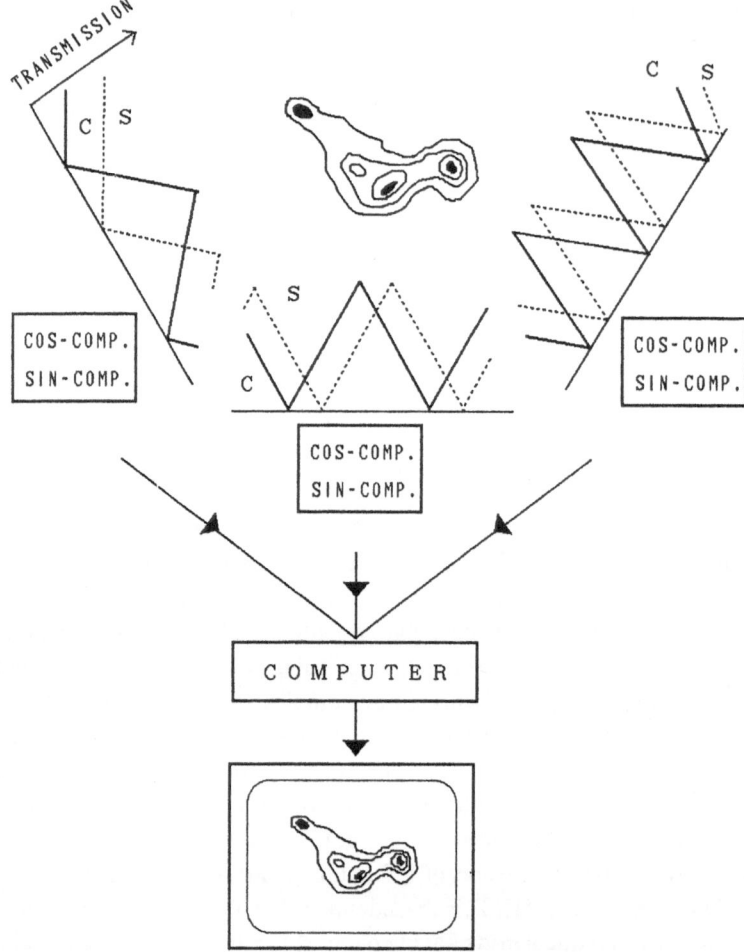

Fig. 2.   Principle of Fourier synthesis for HXT. Pairs of photon counts taken through cosine and sine transmission patterns of subcollimators are synthesized into an image by inverse Fourier transform.

This relation is analogous to that between the cosine and sine functions. The subscript of F (C or S) is added to express this relation. Using the pair of subcollimators, we obtain two photon count data:

$$b_c(k, \theta) = A \int B(X, Y) F_c(k\rho) \, dX \, dY,$$

$$b_s(k, \theta) = A \int B(X, Y) F_s(k\rho) \, dX \, dY,$$

which represent one generalized complex Fourier component. Here $B(X, Y)$ is the X-ray brightness distribution to be imaged and A is an effective area of the subcollimator/ detector system.

If the data actually represented a usual complex Fourier component, and also if measurements were made at a sufficient number of (k, θ)-points, $B(X, Y)$ could be calculated from an inverse Fourier transform. In HXT as in other applications, however, the two conditions are not fully satisfied. In the case of super-synthesis radio telescopes, radio astronomers have developed an image synthesis procedure called CLEAN (Högbom, 1974) for the reconstruction of an image from an incomplete set of measurements. Even the first condition is not a vital one; radio and X-ray astronomers have introduced the maximum entropy method (MEM) for reconstructing an image, where no specific requirement is made on the transmission patterns (e.g., Frieden, 1972; Gull and Daniell, 1978; Willingale, 1981). Thus the sophisticated problems such as the orthogonality and completeness of a system of functions in the mathematical sense are not required to be directly related to the HXT image synthesis problem of ours. Rather our problem is a more practical one.

The following describes our practical approach to image synthesis, using simulated sources and realistic HXT response properties. First we wrote computer programs to simulate the observations and to perform image syntheses. For the image synthesis step, using the analogy of HXT with radio telescopes of Fourier-synthesis type and also benefiting from our previous HINOTORI observations (Tsuneta, 1984; Kosugi and Tsuneta, 1983), we adopted two methods in parallel: one is MEM and the other modified CLEAN. The original CLEAN technique makes use of the Fourier transform to synthesize dirty maps and this transform cannot be applied to our case, so that this part was replaced by a newly-developed procedure.

Next we made numerous computer simulations for a wide variety of model brightness distributions and at different intensity levels. The purposes of the simulations are threefold:

(1) to confirm that the design principle of HXT is correct and that the above-mentioned image synthesis methods work well;

(2) to find the minimum number of subcollimators and also the optimum arrangement of them on the (k, θ)-plane; and

(3) to know the effect of observation errors, namely the effective area error and the phase error, on the quality of synthesized images.

Before making the simulations, a few *a priori* assumptions and restrictions were introduced in the HXT design study. An assumption was that the angular extent of the hard X-ray flare is less than ~ 2 arc min; this determines the fundamental period of repetition or the 'synthesis aperture' ($= 2'06''$). The SMM and HINOTORI results suggest that this assumption is not unreasonable. We chose the highest wave number, k_{max}, to be 8 (in units of the fundamental spatial frequency). This choice is a compromise with the difficulty of fabricating fine grids. Finally we used a symmetrical polar diagram (k, θ), not a rectangular grid for arranging the subcollimators on the (u, v)-plane. This is because an arrangement of subcollimators at regular lattice points on the (u, v)-plane produces on the image plane a repetitive pattern with the fundamental period (so-called grating response) resulting in ambiguity of the flare position.

From the simulations we confirmed the feasibility of partial (u, v)-plane coverage, at least as far as a sufficient number of, say, more than ∼50, complex Fourier components are available. As the number of subcollimators is reduced, the quality of the synthesized images becomes gradually worse. At the same time, the synthesized images from a set of Fourier elements alone tend to deteriorate for relatively extended sources. We found that this tendency can be overcome by replacing the Fourier elements at low wave numbers by what we call 'fanbeam elements' (cf. Figure 3). On the other hand, replacement of Fourier elements at high wave numbers by fanbeam elements results in reduced angular resolution.

Considering all these results together with construction and maintenance problems, we have determined the number of subcollimators to be 64 and chosen as the optimum design the subcollimator arrangement on the (k, θ)-plane as shown in Figure 3.

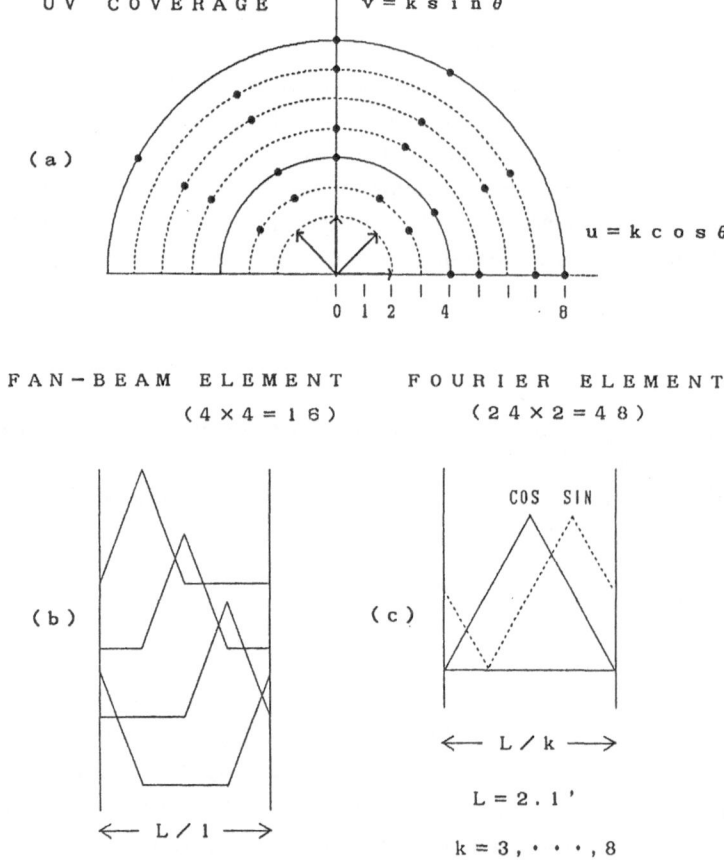

Fig. 3. (a) Arrangement of subcollimators in Fourier-transform space, the (u, v)-plane in which u and v represent the spatial wave numbers of the modulation pattern. Each dot denotes the position at which one pair of Fourier-element subcollimators provides angular information. The fanbeam elements provide information along the arrows near the origin of the (u, v)-plane. The wave number k is given in units of the fundamental repetition period, i.e., 2.1 arc min. (b) The modulation patterns for a set of fanbeam elements. (c) Same as (b) for a Fourier-element pair.

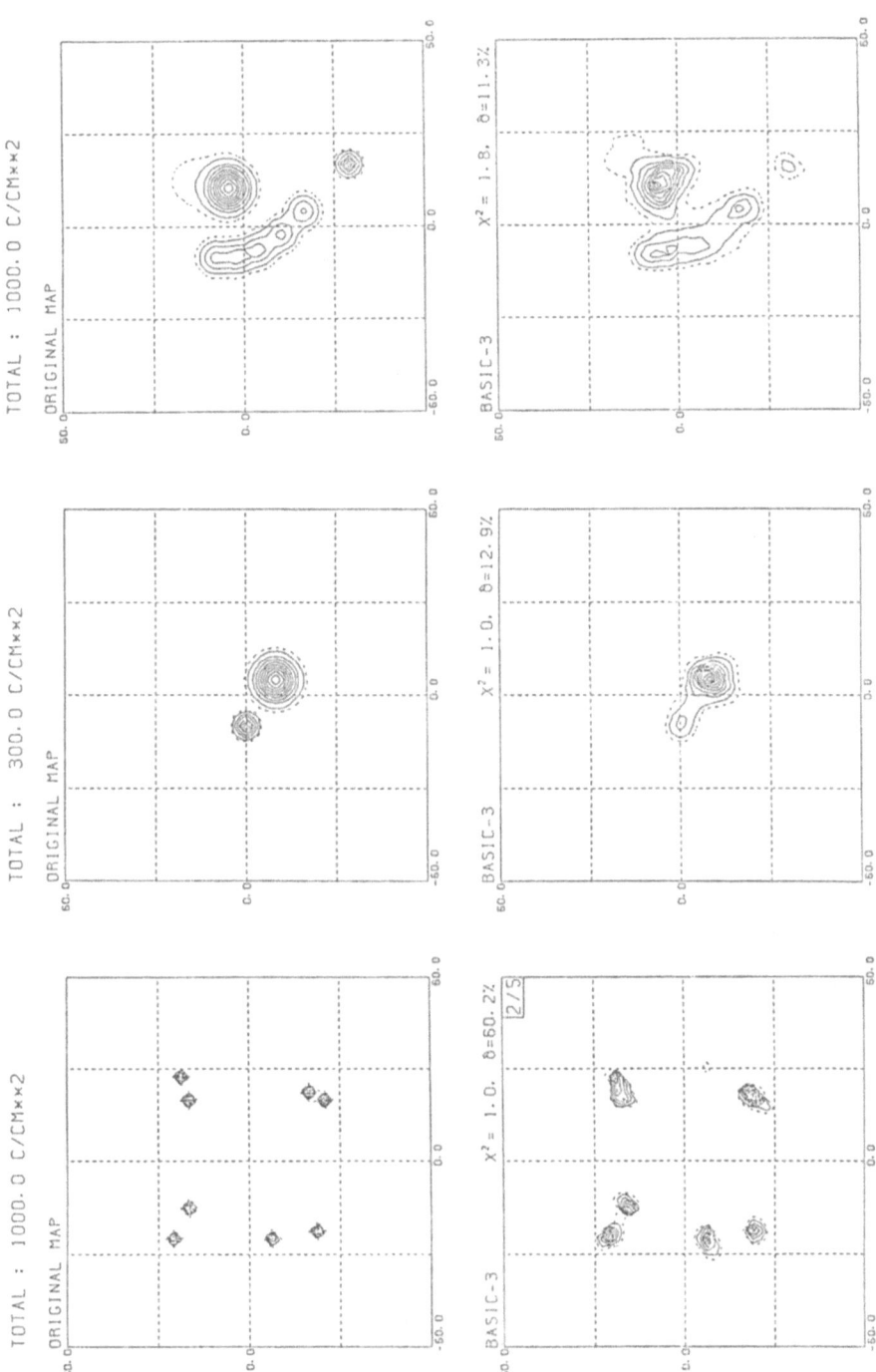

Fig. 4.   Examples of image-synthesis simulations by MEM for the actual HXT design. Restored maps (*bottom*) are compared with the corresponding original maps (*top*) for scattered compact sources (a), double sources (b), and diffuse sources (c). Contour levels are 10% steps (solid lines) and 5% (dashed lines) of the peak. Each box is ~2 × 2′. The incident X-ray fluence is given at the top. The separations between the adjacent sources in (a) are 6″, 8″, 11″, and 15″. The phase error of 1.0″ (r.m.s.) of the 64 transmission functions is incorporated in addition to the Poisson noise in (c).

Examples of synthesized images obtained from the simulations corresponding to the actual design of HXT are shown in Figure 4. The angular resolution experimentally determined from simulations is ~ 5 arc sec.

With regard to the observation errors, we obtained the following results:

(1) The effective area and phase errors do not severely deteriorate the image quality *if* they are correctly evaluated, since they can be removed in the image synthesis procedure.

(2) The synthesized images become quite unstable, when:

– the effective areas of the 64 subcollimators are not evaluated with an accuracy of less than 5% in the r.m.s. sense, or

– the peak positions of the 64 transmission functions are not evaluated with an accuracy of less than 1 arc sec in the r.m.s. sense.

It is to be remarked that the effective area errors are not only related to the subcollimator optics but also to the detector/electronics system of HXT, since the photon count is very sensitive to the energy discrimination levels. Considering the steepness of the flare hard X-ray spectrum (power-law spectrum with the index around $\gamma \approx 4$), the $\sim 5\%$ accuracy in effective area corresponds to $\sim 1\%$ accuracy requirement on the gain of the pulse-height analysis. The peak position error criterion may be slightly relaxed for subcollimators at low wave numbers.

4. The Instrument

From the viewpoint of structure, HXT is divided into three major sections, i.e., the collimator (HXT-C), the detector assembly (HXT-S), and the electronics unit (HXT-E). Functionally it is divided into the X-ray optics part, the X-ray detection/signal processing part, the aspect system (HXA) part, and the power/mode control part. The spacecraft data processor (DP) also plays a role in editing HXT data.

Thus in the following, first we will give an outline of HXT-C, HXT-S, and HXT-E in Sections 4.1 through 4.3, respectively, with concentrated attention to the X-ray detection/signal processing part. The HXA-related items will be described separately in Section 4.4 and the data handling in DP in Section 4.5. The power/mode control of HXT will be mentioned where it is necessary.

4.1. COLLIMATOR (HXT-C)

The roles of HXT-C are the X-ray optics, which provide the 64 independent transmission (modulation) patterns necessary for synthesizing images, and the HXA optics. The two sets of optics require high standards of structural stability to permit coalignment between the individual subcollimators and also between the X-ray and HXA optics systems; note that, if absolute alignment of the order of 1 arc sec is required, a relative shift of one grid with respect to the others smaller than ~ 10 μm should be held. Thus the whole system of HXT-C (except for some minor portions corresponding to two visible-light paths for HXA) is covered by a thermal shield to maintain as homogeneous and stable a temperature as possible. The HXT-C is thermally completely passive.

4.1.1. *Metering Tube*

The metering tube of HXT is a box 1400 mm long and sustains two grid assemblies each 402 mm × 362 mm wide and ~3 kg in weight at both ends by face plates. Because of weight and stiffness requirements, the tube is made of CFRP (carbon-fiber reinforced plastic; $\rho \sim 1.7$ g cm^{-3}) with twelve carbon-fiber sheets stratified quasi-isotropically for minimizing the coefficient of thermal expansion ($\alpha < 1 \times 10^{-6}$ deg^{-1} is achieved). The CFRP box is ~1.3 mm thick at the side plates and ~2.1 mm thick at the face plates, with three stiffeners encircling the side plates. The stiffeners are also used as attachment points to the spacecraft body (cf. Figure 1). The metering tube itself weighs ~8.0 kg.

Mechanically the tube is very resistive against deforming forces. Since any small lateral shift of the rear face plate with respect to the front one can be calibrated by HXA, the most damaging deformation mode for the X-ray optics is a twist between the front and rear plates. The tube stiffness against a twisting torque T is given by

$$\theta \, (\text{arc min}) = 0.025T \, (\text{kg m}) \, .$$

The metering tube is thus extremely stiff, stiffer than the spacecraft center panel to which HXT-C is attached; however, it is still possible that thermal expansion or mechanical deflection of the center panel would deform the metering tube at a certain level. Accordingly a special attachment has been devised to avoid such an effect.

4.1.2. *Grid Assemblies*

The two grid assemblies carry the tungsten grids for the 64 X-ray subcollimators. The aspect lenses with filters are at the center of the front grid assembly, and the fiducial marks at the corresponding position on the rear. The two grid assemblies are fastened by screws to the CFRP face plates.

Each grid assembly consists of a base plate, four fanbeam element units, and six Fourier element units (Figure 5(a)). The base plate is a single molybdenum plate 1.5 mm thick and holds both the X-ray elements and the aspect optics. The plate has 64 square holes (with rounded corners) for the X-ray optics. Each hole has a dimension of 37 mm (front) and 23 mm (rear). The size of the rear hole determines the aperture (effective area) of each subcollimator, and the front hole is larger than the rear one to give full aperture efficiency at any point on the Sun.

Each of the fanbeam element units is a 0.5-mm thick tungsten plate with four fanbeam grids of one position angle, and is set at a corner of the base plate. The grids are fabricated by the electric discharge method. On the other hand, the six Fourier element units, each with eight Fourier grids of one position angle, are the stack of ten tungsten foils, each 50 μm thick, with a molybdenum covering plate 0.5 mm thick. The grids are fabricated by the photo-etching processing. The grid parameters are summarized in Table II.

Figure 5(a) schematically shows the geometrical arrangement of the 64 subcollimators. This diagram shows the slit directions, wave numbers, and the locations of the cosine and sine pairs. The rules for the layout are:

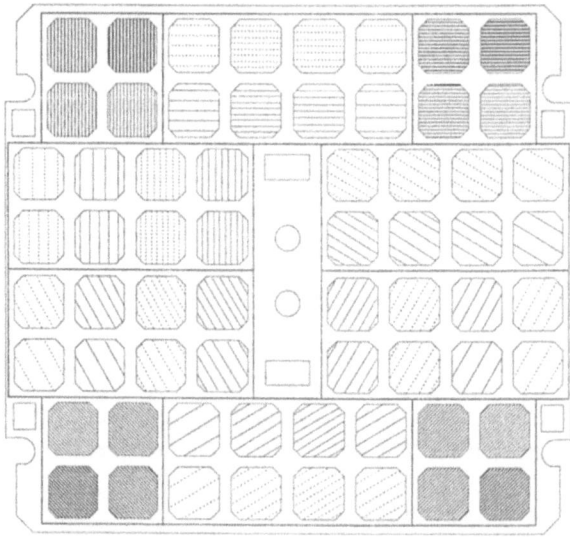

Fig. 5a.   Arrangement of the 64 subcollimators on the grid assemblies (top view). The slit directions of the individual grids are represented by the hatching. The cosine and sine Fourier element pairs are shown by solid and broken hatching lines, respectively, with the number of lines denoting the wave numbers.

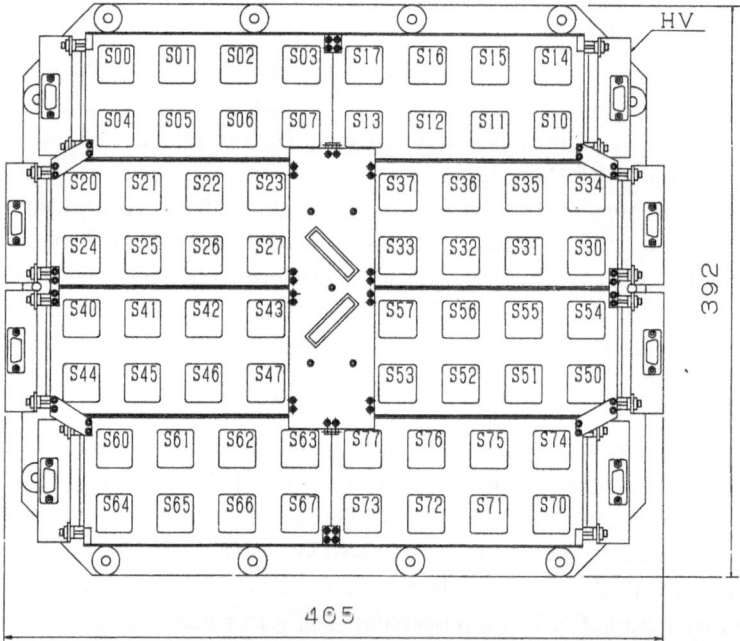

Fig. 5b.   Top view of the detector assembly (HXT-S). The numbered (in octal) squares represent the 64 scintillation counters, grouped in units of 8 modules as shown. At the left- and right-hand sides, the sketch shows the 8 high-voltage supplies, one for each detector unit. The inclined oblong boxes represent the one-dimensional CCD arrays of the HXT aspect system.

TABLE II

Grid main parameters

	Fanbeam elements	Fourier elements
Number of elements	16	48
Mosaic structure	4 el. × 4	8 el. × 6
Position angles	0, 45, 90, 135 deg	0, 30, 60, 90, 120, 150 deg
Number of phases	4 (90° step)	2 (cosine and sine)
Wave numbers	$k = 1, 2$	$k = 3, 4, 5, 6, 7, 8$
Pitch (arc sec)	126	42.0, 31.5, 25.2, 21.0, 18.0, 15.8
Slit width (μm)	210	140, 105, 84, 70, 60, 60
Wire width (μm)	630	140, 105, 84, 70, 60, 45
Material	0.5 mm thick tungsten	0.05 mm thick tungsten foil × 10
Process method	electric discharge	photo etching

(i) The elements of the same position angle are put together in the same unit. This is to make it easy to synthesize one-dimensional images with less danger of phase errors between the same position-angle elements.

(ii) The cosine–sine pair are located adjacent to each other, because to keep the phase difference at exactly 90° is of vital importance.

(iii) Elements with higher wave numbers are concentrated toward the central portions of the assemblies, and the slit directions are selected to be as nearly tangential as possible. This is for reducing a bad effect in case the metering tube happens to twist.

In order to assemble the X-ray optics with very high accuracy, each component has an appropriate number of knockpin holes which ensure a positional accuracy of ~ 5 μm. Also the grid assemblies have six additional grid patterns (two at the center near the aspect optics and four on the fanbeam element units) together with through holes which transmit visible light. They are used for the purpose of coalignment between the front and rear assemblies.

4.2. DETECTOR ASSEMBLY (HXT-S)

As shown in Figure 5(b), the detector assembly is composed of 64 identical detector modules, eight high-voltage power supplies (each for eight detector modules), and two one-dimensional CCD arrays for the aspect optics.

4.2.1. Detector Modules

A detector module consists mainly of a NaI(Tl) scintillation crystal and a photo-multiplier tube; each module has a high-voltage bleeder string and pre-amplifier in its housing (Figure 6). Eight detector modules are packed together to form a detector unit in a magnesium frame, and eight detector units are tied together to form the detector assembly. The detector part is thermally shielded from the high-voltage power supplies, the main heat sources. The high heat conductivity helps to keep the temperature inside the detector assembly homogeneous.

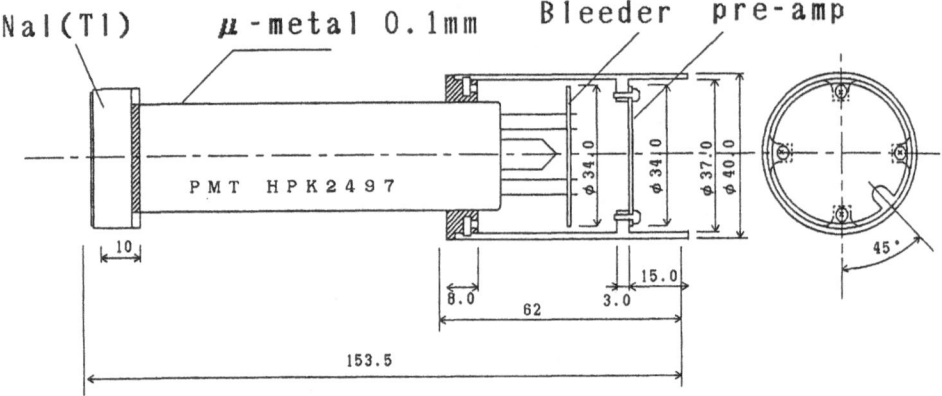

Fig. 6.   Schematic drawing of a detector module.

The crystal has a slightly larger dimension (25 mm) than the grid aperture. It is 5 mm thick, which determines the X-ray detection efficiency at high energies. The crystal is surrounded by an aluminum case. The case, 0.8 mm ($\pm$ 0.01 mm) thick at the front of the crystal, plays the role of an X-ray filter to avoid pulse pile-up due to the flare soft X-rays. In addition to the aluminum filter, X-rays pass through the two CFRP face plates each 2.1 mm thick. Figure 7 shows the spectral response of HXT. (Also is given in this figure the absorption efficiency or the stopping power of the 0.5-mm tungsten plate.)

The photomultiplier tube (HPK 2497 from the Hamamatsu Photonics Co.) is of an anti-vibration type, and is magnetically shielded by μ-metal to suppress gain variation.

The energy resolution (FWHM) of a typical module is given experimentally as

$$\Delta E/E \sim 1.3E^{-1/2} \quad (E \text{ in keV}),$$

a normal value for a NaI(Tl) crystal.

For calibrating the pulse-height gain, a calibration source is attached for each module at the center of the aluminum case front cover. The source is a radioisotope ^{241}Am whose line emission at 59.5 keV is used as a reference point of the pulse-height gain. Since the source is physically small (4 mm^2; $\sim 0.6\%$ of the aperture) and gives only a few to several counts s^{-1} per detector, it does not interfere with flare observations. (A special output mode of HXT data exists for calibration data, see Section 4.3.)

The time constant of the charge amplifier is ~ 10 μs so that each detector can measure incident photon counts up to a few times 10^4 cts s^{-1} without being strongly bothered by the pulse pile-up problem. This limit is large compared to the largest flare yield expected to be $\sim 10^4$ cts s^{-1}.

4.2.2. High-Voltage Power Supply Units

One DC-DC converter supplies high voltage to the eight modules in one detector unit. The output voltage of each converter can be chosen from eight levels by a command from the ground. The eight levels are initially set to be between ~ 800 V and ~ 1050 V

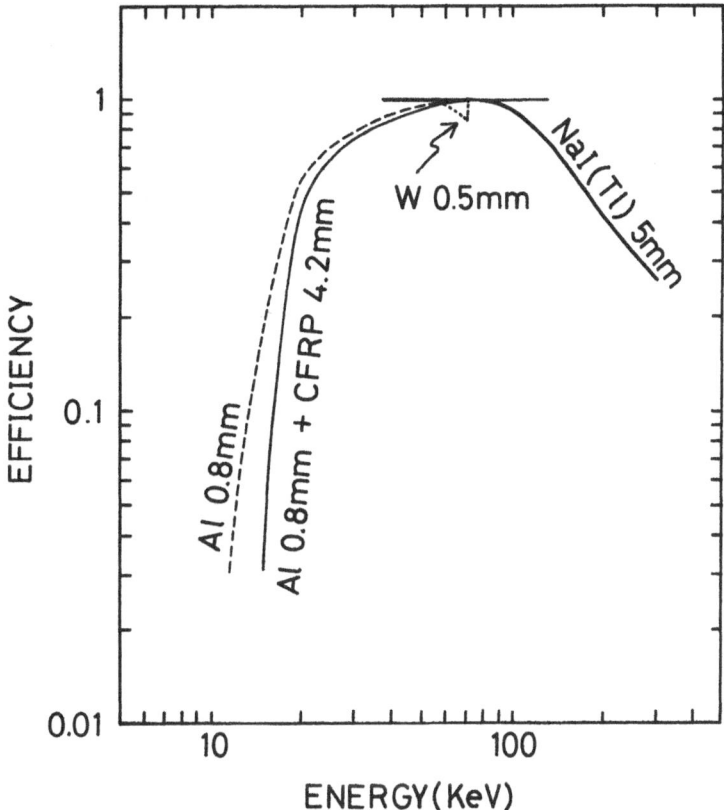

Fig. 7.   Spectral responses of HXT-S and HXT-C. The detection efficiency against X-ray photon energy of the NaI(Tl) scintillation crystal (5 mm thick) is shown by the thick line; the transmission efficiencies of the Al filter (0.8 mm thick) alone and plus the two CFRP face plates (each 2.1 mm thick) by dashed and thin lines, respectively; and the absorption efficiency or the stopping power of the tungsten grids (0.5 mm thick) by dotted line.

with a step of ∼ 35 V. This enables us to adjust the pulse-height gain of the eight modules simultaneously by a step of ∼20% (coarse gain adjustment).

The high-voltage output can be reduced to nearly zero without turning off the control circuit of the supply. This function is used when the spacecraft enters the radiation belt, where photomultiplier tubes could be damaged if high voltage continued to be supplied; leaving the control circuit on increases the stability of the output. This function is also used during satellite night when all the HXT analogue electronics part turns off to reduce the power consumption.

4.3. HARD X-RAY SIGNAL PROCESSING IN THE ELECTRONICS UNIT (HXT-E)

The block diagram of the X-ray signal processes is shown in Figure 8. Each of the analogue signals sent from the 64 modules is first gain-adjusted by an operational amplifier. Here the level of the gain is chosen by a command from 64 levels with a step

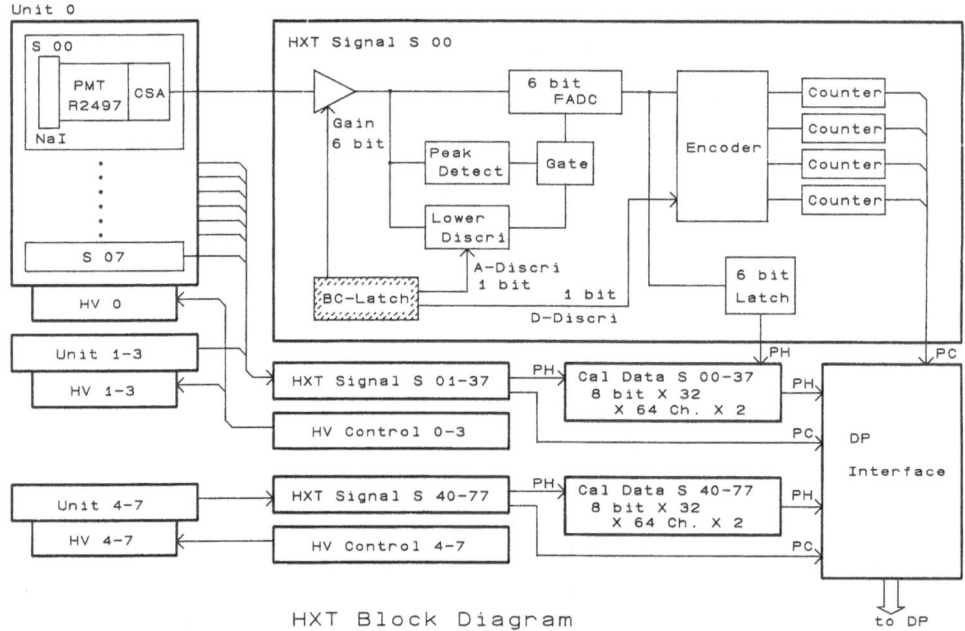

HXT Block Diagram

Fig. 8.   Block diagram of the X-ray signal processing electronics (see text).

of ~1% (fine gain adjustment). Then, in the peak detection circuit, pulses contained in the signal are detected one by one and digitized by a 6-bit flash A/D converter (RCA CA3306D) whenever the pulse height exceeds the analogue discrimination level (selectable from two levels corresponding to ~15 and ~19 keV, both slightly lower than the corresponding digital discrimination levels; see below). The 6-bit value is nominally related to the incident photon energy as

$$\text{energy (keV)} = (1.35 \times \text{value}) + 13.6 \,.$$

Up to this point the 64 modules are independent, each with its own circuit. Further signal processing depends on the operation mode of HXT, observation or calibration.

4.3.1. *Observation Mode (Pulse Count Mode)*

When observing the Sun, each digital data value is sent to an encoder for binning into one of the four counters as detailed in Table III with the nominal photon energy range. The digital value that corresponds to the lower limit of channel 'L' (digital discrimination level) is selectable between values of 1 or 4. Pulses with values lower than this number are not counted in any of the channels.

All photon count data from the 256 (= 4 × 64) counters are sent to the spacecraft data processor every 0.5 s simultaneously. Each counter is 12 bits so that the maximum count is 4095. It can be used cyclically when the photon count exceeds this number, although such big flares rarely occur.

TABLE III

Four PC channels in the observation mode

Channel	Digital data value	Energy range (nominal)
L	1 (4) – 7	15.0 (19.0)–24.4 keV
M1	8–15	24.4–35.2
M2	16–31	35.2–56.8
H	32–63	56.8–100.0

4.3.2. Calibration Mode (Pulse Height Mode)

This mode starts and ends when so commanded from the ground. In this mode each digital value is sent together with the detector module number to the counter memory, where the photon count is binned in 64 channels directly corresponding to the digital pulse heights. The memory consists of two buffers, each of which is 4096 (= 64 × 64)-channel, 8-bit counters, for ensuring data accumulation during the data transfer. These pulse height data are sent to the data processor every 8 s.

Since this mode is used for calibrating the pulse-height gain of the individual modules, it is turned on when the background noise is expected to be the smallest during spacecraft night. In good conditions and with an integration time of ~ 1000 s, we can measure the peak position of the 59.5 keV emission line to an accuracy less than a few percent. Actually the gain calibration measurement will be repeated till a uniformity of the gain to $\sim 1\%$ between the 64 modules is attained. Thanks to the stability of the modules and circuits we expect readjustment to be necessary less than once per month.

4.4. ASPECT SYSTEM (HXA)

The scientific return of HXT strongly depends on whether the positions of hard X-ray sources can be determined with respect to the soft X-ray and visible-light features observed with the Soft X-ray Telescope and ground-based instruments. This is achieved by the aspect system (HXA) with approximately 1 or 2 arc sec accuracy. Although the spacecraft has an attitude control system equipped with aspect sensors (a Sun sensor and a star tracker) and an inertial reference unit which can provide the aspect of the spacecraft with less than 1 arc sec accuracy and a sufficient time resolution, none of them can be coaligned with the X-ray optics of HXT accurately enough for this purpose. The HXA optics, mounted at the center of the two grid assemblies, provide the precise information necessary.

The HXA optics consist of two identical systems. Each is composed of an achromatic, doublet imaging lens 10 mm in diameter with filters on the front grid assembly, a set of fiducial marks located on the rear grid assembly, and a one-dimensional CCD which is placed at the top of HXT-S and measures the white-light brightness distribution of the Sun. The lens center and fiducial marks determine the HXA optical axis, which is referred to as the X-ray 'axis' by metrology. The two CCD's are set to the orthogonal directions to each other.

In HXT-E the CCD video signals are digitized and processed in two complementary ways in parallel. One is that the addresses of pixels whose video signals intersect the discrimination level (selectable from four levels) are sent to the data processor one every second. The other is the output of the complete brightness distributions at a slower rate of once every 64 s (high telemetry bit rate case).

4.5. Onboard Data Handling in the Data Processor (DP)

The spacecraft data processor (Ogawara et al., 1991) receives HXT data and edits them together with other scientific and auxiliary data into a telemetry format according to the observing mode (quiet, flare, night, and BCS-out mode) and the telemetry bit rate (32, 4, or 1 kbps). Here we briefly summarize HXT-related items.

4.5.1. Pre-storage of X-Ray Data for 4 s

The HXT data have no allocation of telemetry words in quiet mode except for channel 'L'. This is not unreasonable because almost no hard X-rays are emitted from outside flaring regions. The problem is, however, that the flare flag can only be set after a flare commences, so that a few seconds elapse before the automatic mode change. In order to avoid loss of initial data of a flare, all the X-ray data from HXT are kept in a buffer for 4 s before sent to the subsequent processing, regardless of mode or the telemetry bit rate.

4.5.2. Time Resolution and Sampling

In flare mode the data processor transmits four-channel HXT data at 0.5 s resolution (high bit rate) or at 4 s resolution after integration (medium bit rate). In addition to this, the channel 'L' data are separately transmitted at 2 s (high bit rate) or 16 s (medium bit rate) resolution as a backup for all observing modes except during spacecraft night when the HXT analogue electronics turn off. Although this backup output is mainly for house-keeping purposes, it also provides scientific data for preflare or postflare hard X-ray images.

4.5.3. Data Compression

Each 12-bit pulse count data word is reduced (after any summing of individual 0.5 s data) to an 8-bit number according to the following function:

$$m = n \qquad\qquad (n = 0\text{--}15),$$
$$m = \text{int}(4 \times \sqrt{n}) \qquad (n = 16\text{--}4080),$$
$$m = 255 \qquad\qquad (n = 4081\text{--}4095),$$

where $\text{int}(x)$ represents the truncated integer value of x. This mapping ensures that digital error due to the bit reduction is always smaller than the Poisson noise statistical error by a factor of two, i.e.,

$$\Delta n' = [(m + 1)^2 - m^2]/16 \sim m/8 \sim \sqrt{n}/2 .$$

Because of this, the n' values estimated from 8-bit m's have almost the same standard deviations as the original n values.

5. Final Remarks

As is summarized in Table I, the design characteristics of HXT greatly improve on those of its two predecessors. Because it is a Fourier-synthesis telescope, it achieves both high sensitivity and wide field of view. The modulation of X-rays up to ~ 100 keV becomes possible owing to the precise fabrication of tungsten grids by the electric discharge and photo-etching methods. The HXT owes its improvement over the HINOTORI hard X-ray imager not only to the precision of its fine grids but also the increased length of the metering tube. Nevertheless, almost all these improvements are simply elements of the design and do not directly determine the practical performance of HXT as an imager. Since HXT is a telescope with 64 independent subcollimators, what determines the image quality of HXT is very complicated. The most important factors are:
  – preciseness of the individual grids,
  – coalignment between the 64 subcollimators within ~ 1 arc sec accuracy,
  – evaluation of the individual patterns as functions of X-ray energy as accurately as possible, and
  – calibration and adjustment of the individual pulse-height gain with accuracy of $\sim 1\%$.

None of these four have been achieved easily. In addition, for hard X-ray images to be successfully compared with images taken with other telescopes, coalignment between the X-ray and HXA optics within a few arc sec is highly desirable.

We have carefully examined the following points during assembling since all the components of HXT were completed in the summer of 1990:

(1) The X-ray grids, examined with microscopes, are well fabricated; individual slits in a grid are cut with accuracy less than 10 μm in position and the average slit pitch differs from the design value by less than 1 μm.

(2) We have examined the coalignment among the 64 subcollimators, using both optical and X-ray methods with the separation between the front and rear grid assemblies reduced, and obtained accurate enough phase-error data for the 64 subcollimators.

(3) The X-ray modulation patterns for the individual subcollimators have been evaluated at photon energies below ~ 40 keV. (No X-ray beams are available at higher photon energies.)

(4) We have checked that the pulse-height gain of the individual detector modules can be adjusted to an accuracy of $\sim 1\%$.

(5) Regarding the coalignment between the X-ray and HXA optics, we have measured the misalignment due to the prism effect, etc., of the HXA optics with an accuracy of a few arc sec.

Calibration measurements will be continuously executed with higher accuracy until the launch.

Acknowledgements

We acknowledge Professors Minoru Oda, Yasuo Tanaka, and the late Prof. Katsuo Tanaka, without whose efforts the SOLAR-A project would not have existed. Dr Keizo Kai, Principal Investigator of HXT and one of the authors of this paper, passed away on 11 March, 1991, one day after this paper was submitted to *Solar Physics*. The rest of the authors wish to express their deep sorrow as well as their will to successfully accomplish this project. The HXT project is a joint effort between the Institute of Space and Astronautical Science, the National Astronomical Observatory, and the University of Tokyo (including both the Department of Physics and the Institute of Astronomy). This work is partially supported by the Scientific Research Fund of the Japanese Ministry of Education, Science and Culture under Grant Nos. 01540214 and 02452011.

References

Dennis, B. R.: 1988, *Solar Phys.* **118**, 49.

Frieden, B. R.: 1972, *J. Opt. Soc. Am.* **62**, 511.

Gull, S. F. and Daniell, G. J.: 1978, *Nature* **272**, 686.

Högbom, J. A.: 1974, *Astron. Astrophys. Suppl.* **15**, 417.

Kosugi, T. and Tsuneta, S.: 1983, *Solar Phys.* **86**, 333.

Makishima, K.: 1982, in Y. Tanaka *et al.* (eds.), *Proc. Hinotori Symp. on Solar Flares*, ISAS, p. 120.

Makishima, K., Miyamoto, S., Murakami, T., Nishimura, J., Oda, M., Ogawara, Y., and Tawara, Y.: 1978, in K. van der Hucht and G. S. Vaiana (eds.), *New Instrumentation for Space Astronomy*, Pergamon Press, New York, p. 277.

Ogawara, Y., Takano, T., Kato, T., Kosugi, T., Tsuneta, S., Watanabe, T., Kondo, I., and Uchida, Y.: 1991, *Solar Phys.* **136**, 1 (this issue).

Prince, T. A., Hurford, G. J., Hudson, H. S., and Crannell, C. J.: 1988, *Solar Phys.* **118**, 269.

Tsuneta, S.: 1984, *Ann. Tokyo Astron. Obs., 2nd Series* **20**, 1.

Tsuneta, S., Acton, L., Bruner, M., Lemen, J., Brown, W., Caravalho, R., Catura, R., Freeland, S., Jurchevich, B., Morrison, M., Ogawara, Y., Hirayama, T., and Owens, J.: 1991, *Solar Phys.* **136**, 37 (this issue).

Van Beek, H., Hoyng, P., Lafleur, B., and Simnett, G. M.: 1980, *Solar Phys.* **65**, 39.

Willingale, R.: 1981, *Monthly Notices Roy. Astron. Soc.* **194**, 359.

Yoshimori, M., Okudaira, K., Hirasima, Y., Igarashi, T., Akasaka, M., Takai, Y., Morimoto, K., Watanabe, T., Ohki, K., Nishimura, J., Yamagami, T., Ogawara, Y., and Kondo, I.: 1991, *Solar Phys.* **136**, 61 (this issue).

THE SOFT X-RAY TELESCOPE FOR THE SOLAR-A MISSION*

S. TSUNETA

Institute of Astronomy, University of Tokyo, Mitaka, Tokyo 181, Japan

L. ACTON, M. BRUNER, J. LEMEN, W. BROWN, R. CARAVALHO, R. CATURA, S. FREELAND, B. JURCEVICH, M. MORRISON

Lockheed Palo Alto Research Laboratory, Palo Alto, CA 94304, U.S.A.

Y. OGAWARA

Institute of Space and Astronomical Sciences, Sagamihara, Japan

T. HIRAYAMA

National Astronomical Observatory of Japan, Mitaka, Tokyo 181, Japan

and

J. OWENS

Marshall Space Flight Center, Huntsville, AL 35812, U.S.A.

Abstract. The Soft X-ray Telescope (SXT) of the SOLAR-A mission is designed to produce X-ray movies of flares with excellent angular and time resolution as well as full-disk X-ray images for general studies. A selection of thin metal filters provide a measure of temperature discrimination and aid in obtaining the wide dynamic range required for solar observing. The co-aligned SXT aspect telescope will yield optical images for aspect reference, white-light flare and sunspot studies, and, possibly, helioseismology. This paper describes the capabilities and characteristics of the SXT for scientific observing.

1. Introduction

The Soft X-ray Telescope (SXT) will provide, for the first time, the opportunity to image the Sun in X-rays over a long period of time with both high temporal and spatial resolution. It gives SOLAR-A an important capability for solar science beyond the study of flares, the primary objective of the mission. The SXT instrument was jointly developed by the Lockheed Palo Alto Research Laboratory and the National Astronomical Observatory of Japan. Collaborators include the University of Tokyo, Stanford University, the University of California at Berkeley, and the University of Hawaii.

The SXT instrument that makes the observations in support of our scientific objectives is a glancing incidence telescope of 1.54 m focal length which forms X-ray images in the 0.25 to 4.0 keV range on a 1024 × 1024 virtual phase charge coupled device (CCD) detector. A selection of thin metallic filters located near the focal plane provides the capability to separate different X-ray energies for plasma temperature diagnostics.

---

* After the launch the name of SOLAR-A has been changed to YOHKOH.

Solar Physics 136: 37–67, 1991.
© 1991 *Kluwer Academic Publishers.*

Knowledge of the location of X-ray images with respect to features observable in visible light is provided by a coaxially mounted visible-light telescope which forms its image on the CCD detector when the thin metallic filter is replaced by an appropriate glass filter.

The ability of the instrument to perform its observational tasks to the levels necessary to achieve our objectives is highly dependent on the optical performance of the X-ray mirror and the quality of the CCD detector. Other determining factors are the stability of the metering structure and the quality of the instrument calibration. Finally, versatility of instrument control and discriminating utilization of limited telemetry are key to the success of the experiment.

1.1. SCIENTIFIC OBJECTIVES

Soft X-ray images reveal the distribution of high-temperature coronal gas and, thus, the structure of the confining magnetic field and thus the topological context of solar activity. SOLAR-A will, for the first time, provide simultaneous soft and hard X-ray images with good angular and temporal resolution. The SXT X-ray images will be searched for the following kinds of information:

 – The geometry of the X-ray emitting structures and the inferred coronal magnetic field topology;
 – the temperature and density of X-ray emitting plasma;
 – the spatial and temporal characteristics of flare energy deposition;
 – the transport of energetic particles and conduction fronts;
 – the presence of waves or other magnetic field disturbances associated with sprays, filament eruptions, and coronal transients; and
 – the locations of energy release and particle acceleration.

The SXT will, by itself, contribute new insights into solar physics. Yet, many studies will benefit from study of correlated observations made with all the SOLAR-A instruments and simultaneous observations made with ground-based solar radio and optical telescopes. Concrete steps have been taken (e.g., Morrison *et al.*, 1991) to facilitate joint analysis of different types of solar observations. The primary objective of the SOLAR-A mission is flare research. SXT will contribute to answering the questions of the following type:

 – Are there observable pre-flare conditions which give rise to an energetic flare?
 – Are there observable discriminators between flares with strong nonthermal effects, e.g., high-energy particle acceleration and mass ejection, and those that exhibit primarily thermal properties?
 – Is flare energy released continuously or in discrete pulses (elementary flares)?
 - What is the filling factor of coronal and flare loops?
 – What is the characteristic time for the acceleration process?
 – Are electrons and ions accelerated simultaneously by the same process? Are there multiple phases or steps in the acceleration process to cover the wide range of energy (non-relativistic to relativistic) and mass (electrons, protons, and heavier ions)?
 – Are there observational clues to the location and dimensions of the acceleration

region? Might it be spatially coincident with the location of hard X-ray and gamma-ray emission?

– How is the energy, generated during a flare, redistributed during evolution through different flare phases?

– Is it possible to observe or infer how and where the energetic particles propagate from the acceleration region? Do they diffuse or propagate in well-collimated beams?

– What is the relationship between the energetic particles which escape from the Sun into the interplanetary space and those which remain at the Sun and produce hard X-ray, gamma-ray, radio, and other emissions?

In addition to the objectives for flare studies the SXT will provide a powerful instrument for non-flare coronal physics. Regular full-disk coronal images will be acquired for the purpose of study of the evolution of the magnetic morphology of the Sun's corona. We hope to elucidate better the appearance, migration, and reconnection of magnetic flux over time scales of a few minutes to a few years by following the creation, change and disappearance of coronal X-ray structures. Studies of the appearance and evolution of coronal magnetic holes will contribute to both solar and solar system physics. We are interested to observe systematically X-ray bright points and clarify their role in the appearance (or disappearance) of magnetic flux. Careful cinematographic studies of the X-ray corona, along with coordinated observations of the visible layers below, may provide new clues to the nature of coronal heating.

The SXT visible-light aspect telescope will provide, in addition to its primary use for precise X-ray/visible alignment information, cinematographic observations of white-light flares and sunspots. Furthermore, a possibility exists to use it for helioseismological studies. It is estimated that intensity oscillations as small as $dI/I \approx 10^{-7}$ may be detected if the SXT data stream is dedicated to this experiment for several months (Sakurai, 1990).

1.2. SXT INSTRUMENT REQUIREMENTS

The SXT has been designed to provide the following capabilities:

– A dynamic range of $> 10^7$ to cover the expected brightness range;

– time resolution of 2 s or better to cover the evolution of the impulsive phase;

– angular resolution of ≤ 3 arc sec to lcoate flare footpoints and observe the filling of loop structures;

– a field-of-view large enough to image the whole solar disk;

– a spectral diagnostic capability sufficient to estimate plasma temperature;

– the capability to record images in visible light, co-aligned with the soft X-ray images, for study of relationships of X-ray and white-light flares and to enable precise registration with other solar data;

– ability to withstand a severe launch environment (20 g r.m.s. vibration levels);

– fundamental mechanical resonance frequency above 100 MHz; and

– the SXT must operate over a temperature range of 0 to $+ 25\ °C$.

2. Optics

In concept, the SXT is a very simple instrument. It has a fixed focus and comprises a sensor, a shutter, dual filter wheels and two co-aligned imaging elements, a mirror for X-rays and a lens for visible light. The position of the filter wheels determine whether an X-ray or an optical picture is taken. There is a commandable door behind the lens, not shown in Figure 1 for clarity, used to exclude visible light from the telescope when desired.

The mechanical and optical design of the SXT required maintaining focus and alignment through a difficult launch and thermal environment with low mass, power,

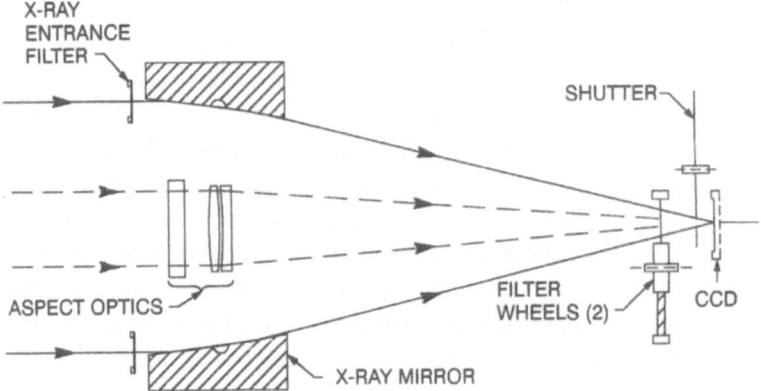

Fig. 1a.   Schematic illustration of the optical concept and key elements of the SXT.

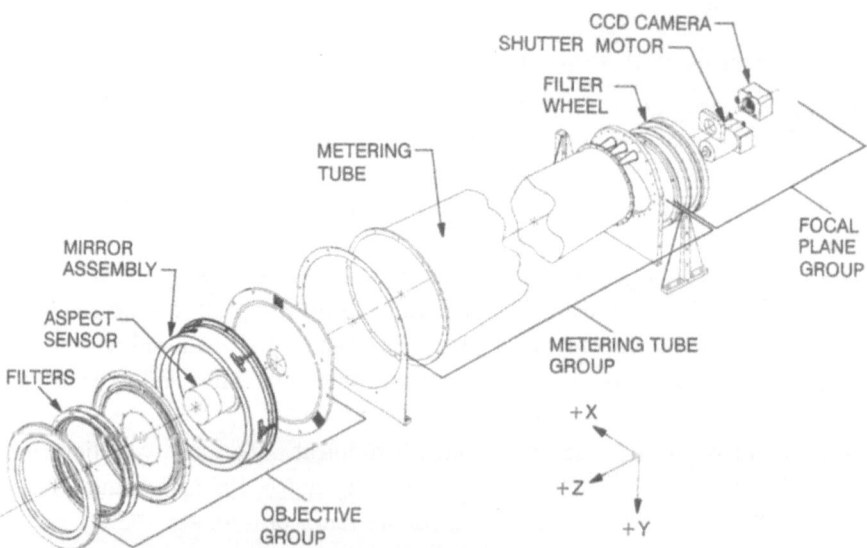

Fig. 1b.   Exploded diagram of the SXT. Sub-assemblies mentioned in the text are identified.

and dollar budgets. The technical solution to these challenges has been described by Bruner *et al.* (1989). The description of SXT in this paper will be limited to that useful for understanding the scientific performance of the instrument. Although the calibration and characterization of SXT given here is preliminary, it is mature enough for confident use in predicting SXT performance and for first-order data interpretation.

2.1. OVERALL SXT RESPONSE

Characteristics of the SXT are presented in Table I. The composite spectral sensitivity of the basic telescope is illustrated in Figure 2(a). Figures 2(b–e) show how this sensitivity is modified by the X-ray analysis filters. These filters have been carefully designed to optimize the ability to determine solar plasma temperature (Section 2.5). The variation of the effective area of the SXT as a function of off-axis angle is illustrated in Figure 3.

The same CCD detector (Section 2.4) is used for X-ray and visible-light images. The metallic X-ray analysis filters or a commandable door block the visible light from the aspect telescope aperture so that it does not contaminate the X-ray images. The angular pixel size of the CCD (2.45″) is approximately the same as the angular resolution of both the X-ray and aspect telescopes. The CCD subtends 42 × 42 min of arc and so furnishes full-disk and coronal coverage in a single image. SXT provides for commandable on-chip pixel summation in order to increase the field of view for the same time resolution. The angular resolution in each summation mode is 1 × 1 (2.45″), 2 × 2 (4.9″), and 4 × 4 (9.8″). The image data are normally compressed and decompressed from 12 to 8 for telemetry and back to 12 bits on the ground for analysis through the use of look-up tables based on a square-root algorithm (Section 3).

Both aspect and X-ray exposures are controlled by a rotating mechanical shutter driven by a constant velocity motor. The shutter blade has two sector openings, one of 3 deg and one of 60 deg. For the very shortest exposure (1 ms) the narrow sector is driven past the CCD without pause. Three such passes in quick succession provide a 3 ms exposure while driving the 60 deg sector past provides a nominal 20 ms exposure. For longer exposures the 60 deg sector is used and the shutter blade stops in the open position for a prescribed period of time. The exposure duration is measured on board by a hardware timer. The shutter is used in combination with a 8.05% transmission metal mesh in the filter wheel to obtain the effective exposures listed in Table II. Note that the exposure error and non-uniformity do not add algebraically. For example, total uncertainty for command 0 will be about 3.0%.

2.2. X-RAY TELESCOPE

Within the SXT, the X-rays are brought to a focus by a glancing incidence mirror of unique technology. This optic utilizes hyperboloids of revolution for both optical surfaces (Nariai, 1987, 1988) to achieve better wide-field angular resolution on a flat focal plane than would the familiar paraboloid-hyperboloid design. Wide-field performance is further enhanced by making the mirror unusually short (4.5 cm total) along the optical axis (Watanabe, 1987). Both optical surfaces of the mirror are formed in a single cylinder

TABLE I

SXT characteristics

General	
Weight	
telescope	14.7 kg
electronics	9.0 kg
Average daytime power	7–12 W, plus heaters
Envelope	30 × 30 × 170 cm
Control	On-board computers
CCD detector (operated cooled to − 18 °C)	
Array size	1024 × 1024
Pixel size	18.281 ± 0.002 μm = 2.4528 ± 0.0005″
On-chip summation	2 × 2 and 4 × 4
Digital signal	12 bit compressed to 8 bit
Time resolution	0.5 s (special mode)
64 × 64 pixel image	2.0 s (normal)
X-ray telescope	
Diameter	230.65 mm
Geometrical area	261.75 mm^2
Peak effective area	78 mm^2 at 8 Å
Effective focal length	1535.6 mm
Spectral range	3–45 Å at 1% of peak response
Resolution	≤ 3″ FWHM over solar disk at 8 Å
Plate scale	134.2″ mm^{-1}
Dynamic range	> 5 × 10^9
Aspect telescope	
Aperture	50 mm
Effective focal length	1538.4 ± 0.5 mm
Transmission	0.013% from 3500 to 4500 Å
Filter wheels	

Commanded position	Front wheel	Rear wheel
1	Open	Open
2	30 Å at 4310 Å	Al 1265 Å
3	CCD flood lens	Al/Mg/Mn composite
4	Opal-glass diffuser	Be 119 μm
5	140 Å at 4580 Å	Al 11.6 μm
6	8.05% mesh	Mg 2.52 μm

Rotating sector shutter	
Effective exposure range	0.077 ms–242 s
Number of steps	37

of low-expansion Zerodur glass-ceramic which is bonded into a lightweight titanium stress-free mount. The reflecting surfaces are covered with an evaporated coating of 420 Å of gold on top of 80 Å of chromium. For wavelengths longer than 6 Å where scattering is moderate the agreement between predicted and calibrated mirror effective area agree to better than 90%. Further description of this optic and its calibration is published separately (Lemen *et al.*, 1989, 1991).

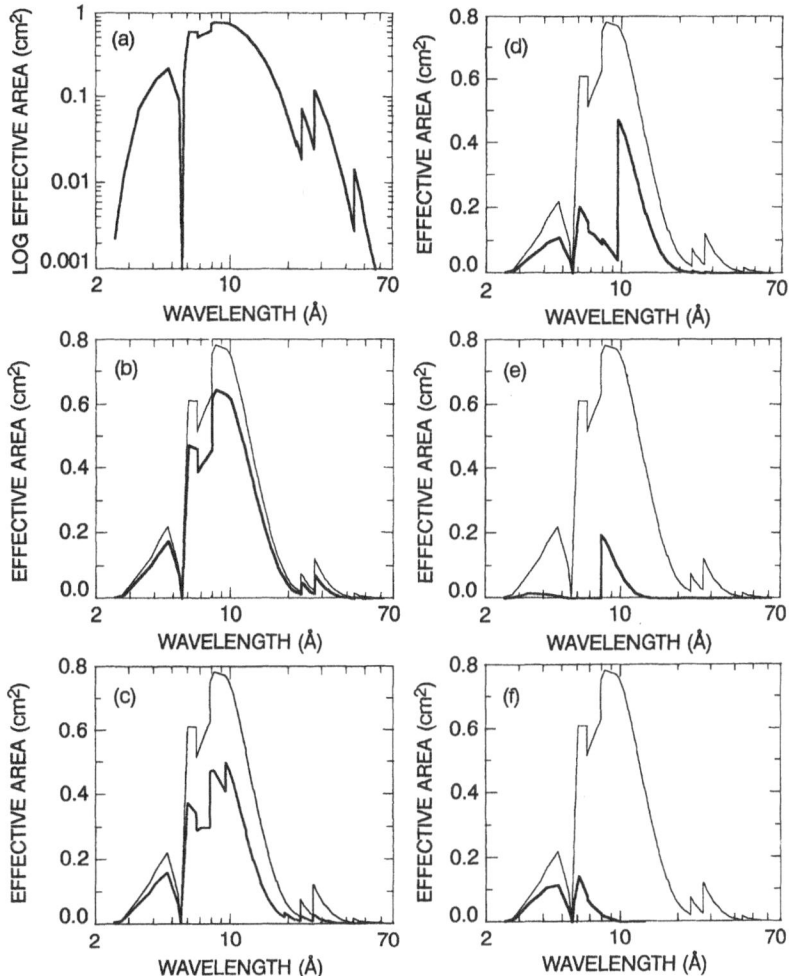

Fig. 2. Effective area of the SXT: (a) no analysis filter, (b) 1265 Å Al filter, (c) composite filter comprising 2930 Å Al, 2070 Å Mg, 562 Å Mn, and 190 Å C, (d) 2.52 μm Mg, (e) 11.6 μm Al, and (f) 119 μm Be. (b)–(f) compare the no-filter case (light line) with the filtered case (heavy line). (a) is plotted on a logarithmic scale to better illustrate regions of small effective area, which may dominate SXT response when the solar spectrum is intense at those wavelengths.

Glancing incidence X-ray mirrors are often troubled by broad, intense, scattering haloes. This is especially troublesome for solar flare observing where scene contrast over small angular scales is extreme and, at the same time, scientifically important. This small angle scattering is dominated by surface roughness and mid-frequency errors of figure. For SXT, a surface roughness of 3.8 Å r.m.s. and mid-frequency error of 51 Å r.m.s. was achieved. The point spread profile of the telescope, acquired with a flight-type CCD, is illustrated in Figure 4.

The angular resolving power of an under-sampled X-ray telescope like SXT is not

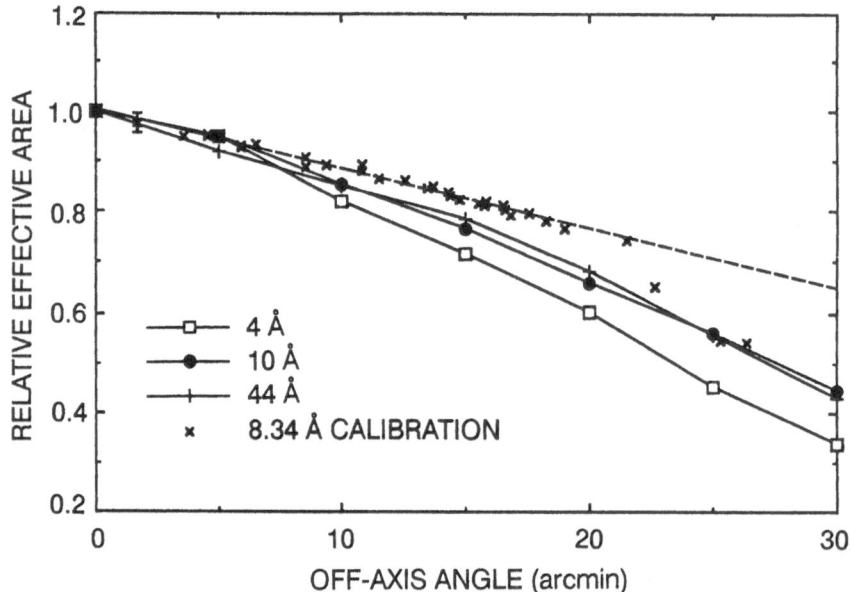

Fig. 3.   Variation of effective area of SXT with off-axis angle determined by ray tracing (without X-ray scattering) for three different wavelengths. Calibration data at 8.34 Å are shown with a straight line fit to the calibration results inside of 23 arc min.

easy to characterize in simple terms. The X-ray point spread function (PSF) of the instrument is defined as the empirical expression that describes the intensity distribution over the image of a point source at infinity. The PSF is a function of wavelength and off-axis angle. One figure of merit useful for X-ray mirrors is the diameter, D_{50}, of the circle that encloses 50% of the imaged energy. For SXT, calibration data yields the following expression for this parameter (on-axis) over the 4–45 Å interval:

$$D_{50} = 7.0 - 2.4 \log_{10} \lambda \text{ arc sec} , \qquad (1)$$

where λ is the wavelength in Å.

A second common means of characterizing the PSF is the full width at one-half maximum (FWHM) of the best fitting 2-dimensional Gaussian (Figure 5). Quantitatively, this is a poor choice for comparing telescopes because the intrinsic PSF has a very sharp central spike (Lemen et al., 1989). For such a PSF the FWHM is sensor-dependent because the definition of maximum reflects pixel size down to very small pixels (or film grain). Also, the wings of the profile are distinctly non-Gaussian. A modified Moffat PSF (Bendinelli, 1991) has been found to fairly accurately characterize the PSF of the SXT. The form of this expression is

$$N = \frac{C}{\left[1 + \left(\dfrac{r}{a} \right)^2 \right]^b} , \qquad (2)$$

TABLE II

SXT effective exposures

Command	Effective exposure	Error (%)	Non-uniformity (%)
0	0.077 ms [a]	1.4	2.6
1	0.23 ms [a]	1.4	2.6
2	0.96 ms	1.4	2.6
3	1.38 ms [a]	1.4	2.3
4	2.88 ms	1.4	2.6
5	3.08 ms [a]	1.3	≤0.1
6	4.69 ms [a]	0.8	≤0.1
7	6.30 ms [a]	0.6	≤0.1
8	9.50 ms [a]	0.4	≤0.1
9	17.2 ms	1.4	2.3
10	28.8 ms	3.2	1.0
11	38.2 ms	1.3	≤0.1
12	58.3 ms	0.8	≤0.1
13	78.3 ms	0.6	≤0.1
14	118 ms	0.4	≤0.1
15	168 ms	≤0.4	≤0.1
16	238 ms	≤0.4	≤0.1
17	338 ms	≤0.4	≤0.1
18	468 ms	≤0.4	≤0.1
19	668 ms	≤0.4	≤0.1
20	948 ms	≤0.4	≤0.1
21	1.34 s	≤0.4	≤0.1
22	1.89 s	≤0.4	≤0.1
23	2.67 s	≤0.4	≤0.1
24	3.78 s	≤0.4	≤0.1
25	5.34 s	≤0.4	≤0.1
26	7.55 s	≤0.4	≤0.1
27	10.7 s	≤0.4	≤0.1
28	15.1 s	≤0.4	≤0.1
29	21.4 s	≤0.4	≤0.1
30	30.2 s	≤0.4	≤0.1
31	42.7 s	≤0.4	≤0.1
32	60.4 s	≤0.4	≤0.1
33	85.4 s	≤0.4	≤0.1
34	121 s	≤0.4	≤0.1
35	171 s	≤0.4	≤0.1
36	242 s	≤0.4	≤0.1

[a] The 8.05% transmission mask is applied.

where N is the signal in a pixel in digital data numbers (DN), C, a, and b are parameters of the fit, and r is the radial coordinate of the function, expressed in pixels.

The SXT is equipped with two entrance filters in series for the purpose of excluding visible and UV light from the telescope. The filters comprise Lexan as the support material, Al for thermal control and to exclude visible light, and Ti to exclude the intense solar He 304 Å emission. The dual filter design was chosen to minimize the possibility of stray light entering the instrument through a pinhole in the metal films.

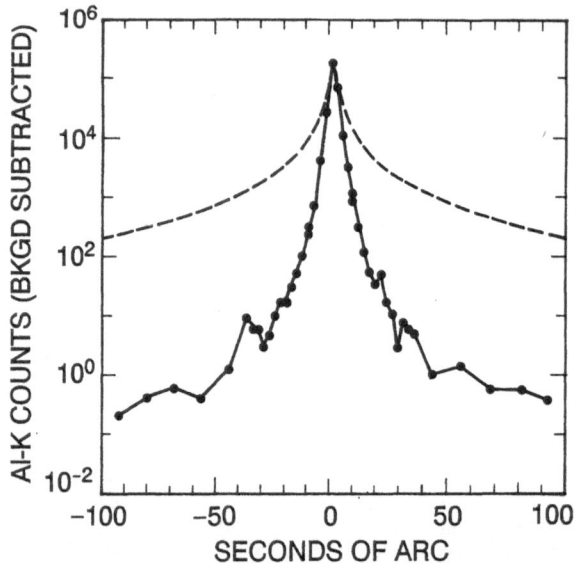

Fig. 4. Measured point spread profile of SXT at 8.34 Å (Al K line) for 1″ source at 310 m. Equivalent curve for Skylab SO-54 telescope (Vaiana *et al.*, 1977) is shown as a dashed line for comparison. Great improvement in decreased scattering wings for SXT is evident.

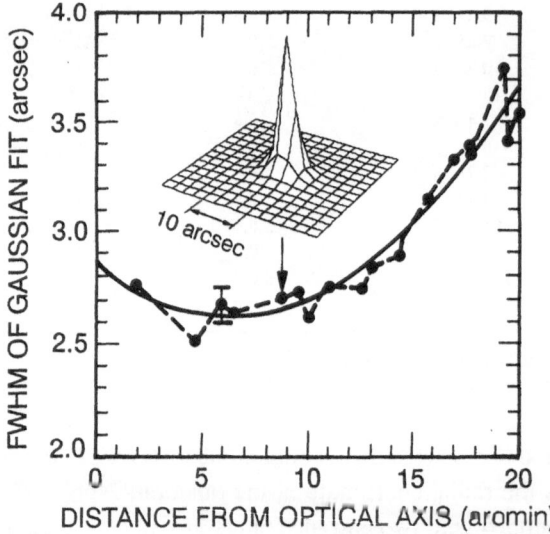

Fig. 5. Approximate FWHM of the SXT image of a point source versus distance from optical axis. The vertical scale is seconds of arc with an estimated uncertainty of about 10%. The inset shows the surface plot of the image at the point indicated by the arrow. A typical 1σ error bar is shown. A parabola has been fitted through the points. The CCD has been deliberately placed 0.1 mm ahead of the position of best on-axis focus in order to achieve a more uniform resolution across the solar disk.

The 5 X-ray analysis filters of the SXT, located in the rear filter wheel, are designed to provide attenuation, increasing the dynamic range of the instrument, and a measure of flare temperature information (Section 2.5). Four of the filters are supported on stainless steel mesh. Figure 2 illustrates the response of the SXT through the various filters. The 8.05% transmission stainless steel mesh located in the front filter wheel can be used in tandem with any of the positions of the rear wheel to extend the dynamic range of the instrument. Analysis of calibration images has not revealed any measurable effect on images taken with this attenuating mesh.

2.3. ASPECT TELESCOPE

The SXT aspect sensor is a high-quality, albeit small, optical telescope in its own right. It serves several purposes:

(1) Sunspot and limb images for determining SXT pointing to an accuracy of 1 sec of arc or better and to aid image registration with ground-based data.

(2) To record magnetic plage, sunspot and pore motions and development in and around active regions.

(3) To observe white-light flares.

(4) For helioseismology.

(5) To provide flat field illumination for CCD gain calibration.

(6) To provide a source of blue light for photon flood of the CCD for the purpose of annealing soft X-ray degradation of the sensor.

The objective lens assembly forms an image of the same scale as the X-ray image that is co-aligned to approximately one pixel. The actual co-alignment will be verified in orbit by overlay of the white-light limb with the X-ray absorption limb using full-disk images. The assembly allows just enough light into the SXT to acquire properly exposed aspect pictures in about 0.1 s.

The entrance filter consists of a white-light attenuator and a bandpass filter. The white-light attenuator is composed of a synthetic fused silica substrate with a 500 Å aluminum attenuating layer, under a dielectric coating for durability, on the rear surface. The bandpass filter is deposited on a 6 mm thick substrate of Hoya CM-500 blue glass. The transmission of the white-light attenuator and the bandpass filter in tandem is illustrated in Figure 6. The combined out of band rejection of the white-light attenuator and bandpass filter is of order 10^{-8}.

The image of the aspect telescope is formed by a doublet lens, which is achromatic across the entrance filter's passband. The lens has a 50 mm clear aperture and a focal length of 1538 mm, forming a beam which is approximately $f/31$. The doublet has a depth of focus of approximately ± 0.5 mm, with an Airy disk diameter of about 50 microns. Lenses have been selected to give the best match to the measured effective focal length of the SXT X-ray mirror. Optical materials for the objective lens assembly have been selected, by specification or test, for their radiation resistance. All appropriate surfaces have been anti-reflection coated.

The SXT filter wheel carries two optical band pass filters. Magnetic plage in and around active regions will be readily visible in the 30 Å filter centered on the CN band

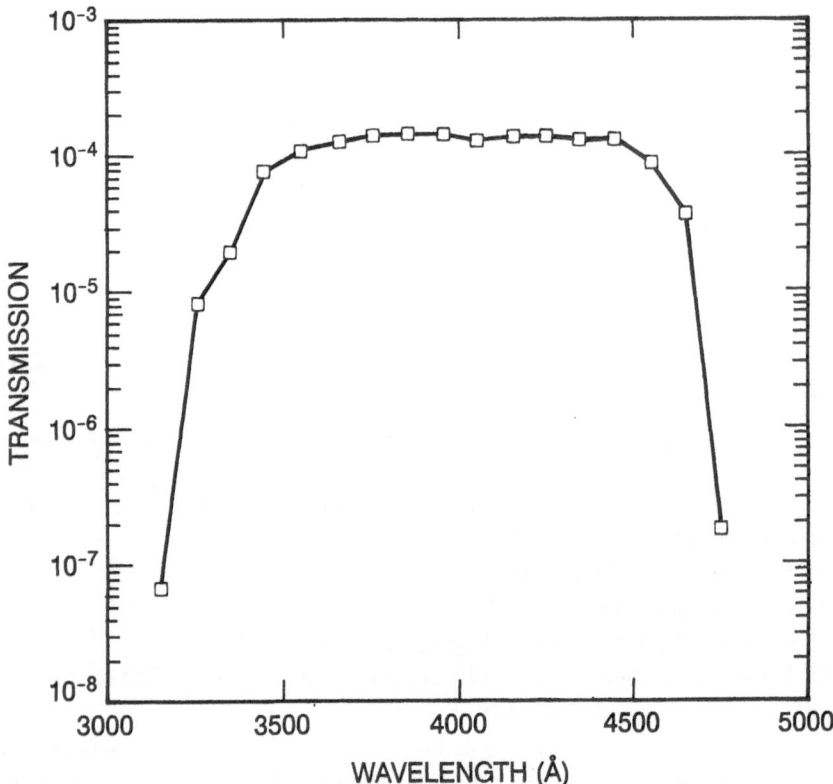

Fig. 6.   Spectral transmission of the objective lens assembly.

at 4308 Å. In order to acquire a continuum image with a filter that is less subject to drift over the life of the mission we have also included a 140 Å filter centered at about 4580 Å. These filters are equipped with neutral density attenuators on the rear side to balance the exposure times and to attenuate ghosts from back reflections off of the CCD to less than 1%. The bandpasses of these two filters are illustrated in Figure 7.

2.4. CCD IMAGE SENSOR

The CCD camera for the SXT utilizes a 1024 × 1024 virtual phase CCD (VPCCD) with 18.3 μm pixels, manufactured specifically for SXT by Texas Instruments at their Miho, Japan, facility. The principal of operation of the VPCCD has been described by Hynecek (1979) and Janesick, Hynecek, and Blouke (1981). Virtual phase technology is a good choice for SXT because the thin oxide layer covering the virtual well provides excellent soft X-ray response without the difficulties of thinned, back illuminated, operation.

The SXT camera is operated in a charge-collection, rather than a photon counting, mode. Solar features are typically bright enough to produce near full well images with exposures of less than 1 s. Because of the high signal levels, SXT requirements on read

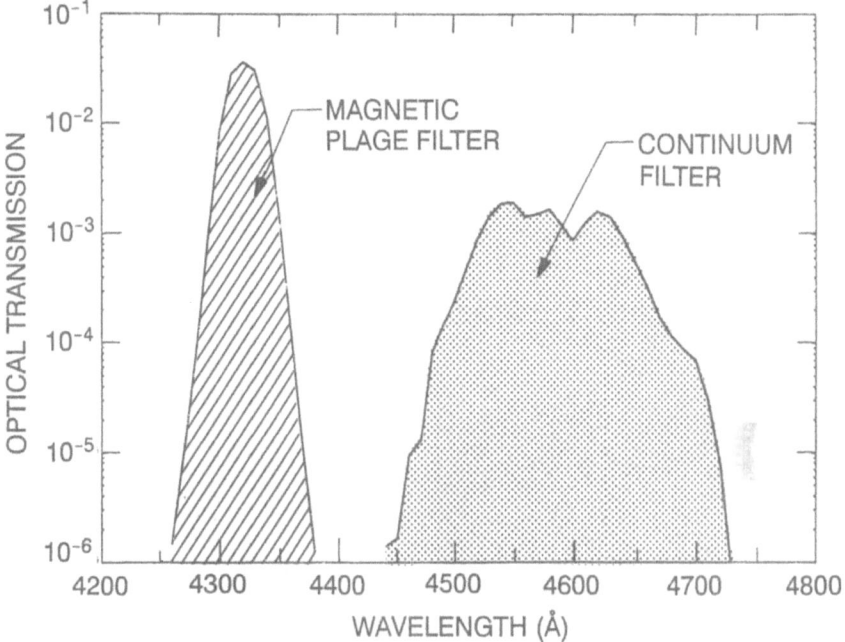

Fig. 7.   Spectral transmission of the aspect telescope bandpass filters.

noise and dark current are not stringent. Dark-current control and gain stability are obtained by cooling the CCD to $-18\,°C$ with a closed-loop 3-stage thermoelectric cooler. CCD read noise is about 85 electrons pixel^{-1} r.m.s. Average dark current at $-18\,°C$ is about 9 electrons pixel^{-1} s^{-1} (0.4 picoamp cm^{-2}). Charge transfer efficiency of the SXT flight sensor measured 0.999989 for signals $> 10^4$ electrons. The forward filter wheel includes an opal glass diffuser to provide flat field illumination of the CCD for the purpose of gain calibration through use of the photon transfer curve technique (Janesick, Klaasen, and Elliott, 1987). Figure 8 shows the pixel structure and the measured quantum efficiency of the SXT flight device. The quantum efficiency of the CCD for blue light from the aspect sensor is roughly 30%.

The full well capacity of the CCD is about 250 000 electrons. At the conversion constant of 3.65 eV electron^{-1} for Si this results in a full well capacity of about $10^3 \times 1$ keV (12.4 Å) photons. Extensive use of 2×2 and 4×4 on-chip summing is anticipated so the camera gain has been set low (100 electrons per digital number (DN)) to preclude excessive saturation of the 12 bit analog to digital convertor. For full-resolution observations (no on-chip summation) the CCD will saturate before the 12-bit ADC attains full scale in order to assure that maximum CCD signal capacity is available. For summed exposures the ADC will saturate first. Managing to achieve correct exposures, especially on flares, will be challenging. This task will normally be assigned to the on-board automatic exposure control described below. The CCD

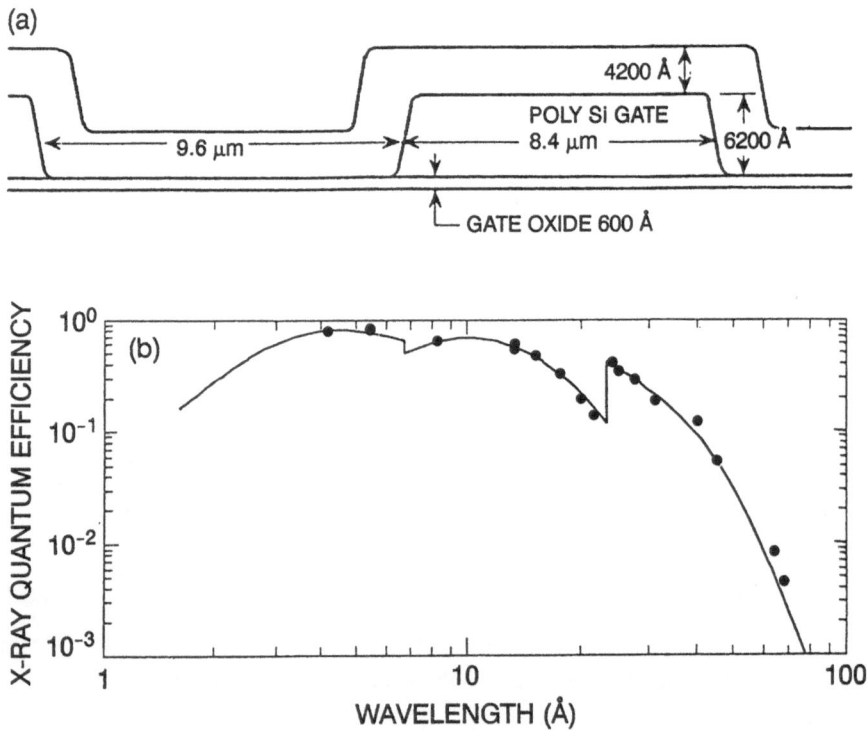

Fig. 8.   (a) Pixel structure of CCD from electron microscopy. (b) Solid curve shows the computed quantum efficiency (QE) of the CCD calculated from the measured pixel structure. The points are QE calibration data obtained at Lockheed for the SXT flight device. Statistical error of these data is smaller than the plotted points.

camera can also be operated in a pseudo-frame transfer mode in the event of problems with the mechanical shutter.

The SXT CCD is subject to degradation from cosmic rays, energetic protons trapped in the Earth's radiation belts and from the solar X-rays themselves. These problems have been treated in detail for SXT by Acton *et al.* (1991). Because the sensor is operated cooled and exposures are normally a fraction of a second, the increased dark current and dark spikes (pixels with greatly enhanced dark current) produced by energetic particles do not appear to pose a significant threat to SXT science over the 3–5 year mission lifetime. However, ionization produced in the thin oxide layer of the CCD by the solar X-rays produces increased dark current and, eventually, loss of charge transfer efficiency with resulting smearing of the solar images. It has been discovered that illumination of the CCD with blue light from the aspect telescope apparently creates free electrons at the Si–SiO$_2$ interface, neutralizing the charge and annealing the damage. The SXT filter wheel has been equipped with a slightly negative quartz lens for the purpose of flooding the entire CCD with 3300–4700 Å light. The duration of this photon flood can be set by command. Present plans call for doing such a flood for approximately the first 4 min of each daylight pass, while the spacecraft's attitude is stabilizing.

2.5. SXT RESPONSE TO THE SUN

The SXT is capable of imaging solar plasma over the temperature range of < 1 to > 50 MK for a wide range of intensities. The sensitivity of the SXT for detecting solar features of different temperatures is illustrated in Figure 9. These curves are produced

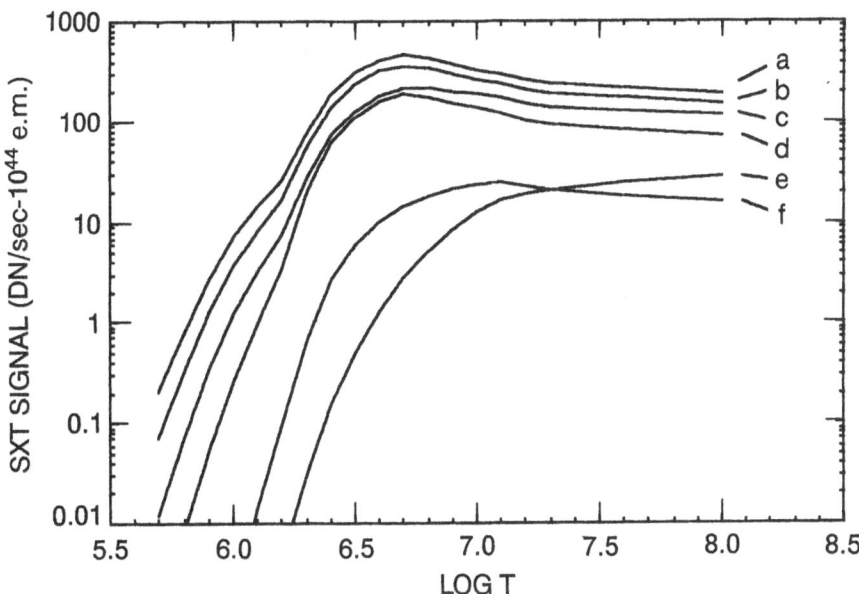

Fig. 9.   Total signal as a function of $\log_{10} T$ for the open filter case (a) and the SXT analysis filters: (b) Al 1265 Å, (c) Al/Mg/Mn, (d) Mg 2.52 μm, (e) Al 11.6 μm, (f) Be 119 μm.

by convolving the instrument response functions of Figure 2 with the theoretical X-ray emission line spectra of Mewe, Gronenschild, and van den Oord (1985) and the continuum expression of Mewe, Lemen, and van den Oord (1986). Note that these curves assume an emission measure of 10^{-44} cm^{-3}. In thinking of SXT sensitivity it is important to keep in mind that, even for a point source, only about half of the CCD signal will appear in the brightest pixel.

Because the solar soft X-ray spectrum is predominantly a line spectrum, the well-defined absorption edges (Figure 2) of the SXT analysis filters emphasize detection of certain spectral lines or groups of lines. This provides rudimentary spectral, and hence temperature, discrimination. Figure 10 presents the ratio of signals versus $\log_{10} T$ for selected pairs of analysis filters. The two thickest filters will, in general, not work for non-flare temperatures because of inadequate sensitivity. Similarly, the open telescope (no analysis filter) and thinnest filters may saturate in flares, even for the shortest exposures. For isothermal sources the uncertainty in plasma temperature determined by this technique is of about 0.1 in $\log_{10} T$.

In principle, the response curves of Figure 9 can be treated in the same way as the emissivity curves for an atomic emission line for differential emission measure (DEM)

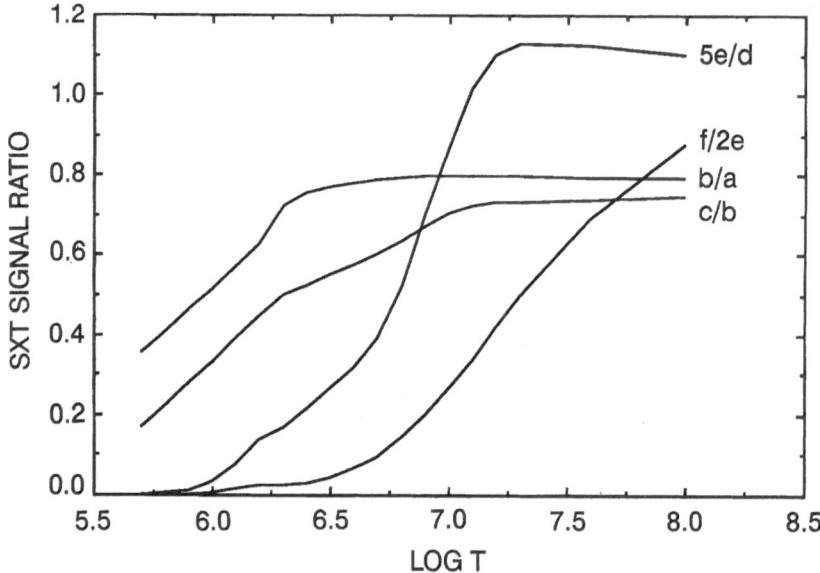

Fig. 10.   Ratios of the SXT response functions of Figure 9. In 2 cases, 5e/d and f/2e, observing time with e, the 11.6 μm Al filter, must be increased by factors of 5 and 2, respectively, with respect to the other filter to achieve the given ratio.

analysis of a multi-thermal plasma. In practice, the problem is more ill-conditioned (Craig and Brown, 1986) than the atomic line case. Figure 11 illustrates the results of a simulation of such an analysis. Here, SXT signals were predicted for the given DEM and put through a version of the Sylwester analysis program (Fludra and Sylwester, 1986) to generate a DEM (Strong *et al.*, 1991). This work indicates that the DEM reliably reflects the presence of high-temperature plasma and the total emission measure is well determined. However, detailed structure or peaks in the DEM distribution are not reproduced and the low-temperature end is poorly determined in the presence of Poisson counting statistics.

Photon statistics will contribute the dominant error in SXT photometry. A CCD signal of 100 electrons produced an increment of one digital number (DN) from the ADC. At 3.65 eV electron^{-1}, this corresponds to detecting a single photon of wavelength 34 Å. The 1σ readout and detector noise on a typical image will be roughly 1 DN. The true photon statistics in a thin-filter image will be additionally uncertain because of the factor of 10 range in photon energies to which the SXT is sensitive. Thicker filters restrict the spectral acceptance band of the instrument so that the statistical error can be more accurately estimated.

Table III demonstrates the predicted signal from a single SXT pixel through each of the SXT analysis filters for a variety of solar features (Strong and Lemen, 1987) and exposures appropriate for their observation. The angular extent of typical features is taken into account. McTiernen (1991) has examined the response of the SXT to two non-thermal (power-law spectrum) events which were partially occulted by the Sun from

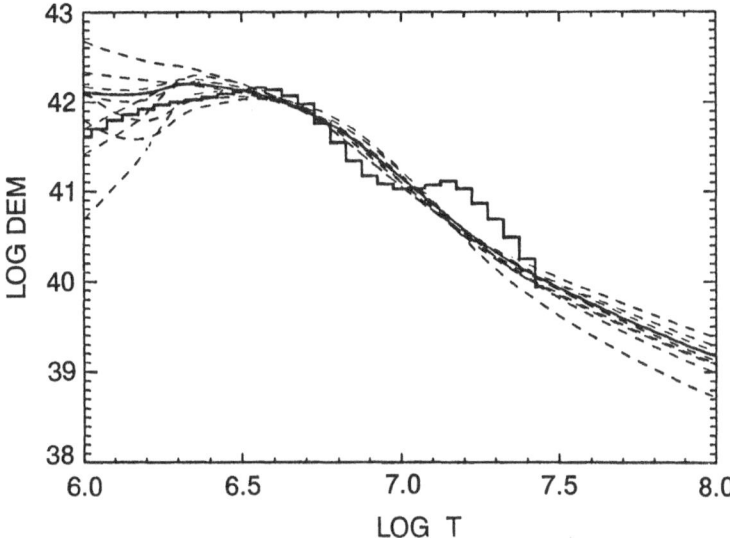

Fig. 11. Results of a simulated derivation of a differential emission measure distribution (DEM) using SXT images through all 6 analysis filters (Strong *et al.*, 1991). The histogram is the input distribution. The solid curve is the best-fit model derived from the input data. The dashed curves represent additional random solutions reflecting Poisson counting statistics.

the ISEE-3/ICE spacecraft (Kane *et al.*, 1979). He estimated that the flare of 16 February, 1984, which was visible beyond 160 000 km above the surface of the Sun and exhibited a photon power-law spectral index of about 4.6, would have generated a signal of 500–1500 DN pixel^{-1} s^{-1} in the SXT with no filter. Such sources in the high corona above the limb should be observable by SXT. The spectral index of the emission can be found using thick/thin filter ratios. For this flare the uncertainty in the spectral index using a ratio of the thick beryllium filter to the thin aluminum filter would be approximately 0.6, although the SXT alone could not distinguish between power law and thermal spectra.

3. SXT Image Data

The acquisition of solar images by the SXT is controlled by electornics and software operating within the SOLAR-A Data Processor (DP). This function is slaved in frequency to the SOLAR-A telemetry rate. Images will normally be taken under either high (32 kbps) or medium (4 kbps) rate. Low rate (1 kbps) is used only for night mode. In the discussion to follow, examples of image frequency will always assume high-rate telemetry unless otherwise noted. For medium rate operation, picture frequency decreases by a factor of eight.

The CCD camera employs a 12-bit ADC but data are transferred to the DP as 8-bit numbers. For purposes of gain calibration and helioseismology it is important to maintain the full 12-bit accuracy. This is accomplished by transferring either the high-

TABLE III

Examples of SXT response

Feature	Temp. (MK)	Em. meas. (cm^{-3})	Exposure (s)	Open	Al 1265	Al/Mg/Mn Signal (DN) in one pixel	Mg 2.5	Al 11.6	Be 119
Coronal hole	1.3	2E+42	60.4	18	10	4	1	0	0
X-ray bright point	1.8	7E+43	0.948	34	23	11	6	0	0
Large-scale loops	2.1	6E+42	5.34	30	22	11	8	0	0
Active region	2.5	3E+45	0.468	sat.	2034	1064	878	37	2
Long duration event (LDE)	7.5	1E+48	0.00096	sat.	sat.	1958	1515	201	74
Impulsive C flare	12	5E+48	0.00023	sat.	sat.	2047	1449	286	180
M flare	17	2E+48	0.00023	1191	938	690	464	101	92
X flare	20	1E+49	0.000077	1886	1501	1101	723	162	163
Post-flare loop	7	3E+46	0.0172	2084	1671	1070	856	104	344

order or low-order 8 bits, selectable by command. Normally, the ADC numbers are compressed to 8 bits according to the following algorithm. Taking, N, X, and M to be the original, compressed and decompressed digital numbers, respectively,

 – for $N \leq 64$:

$$X(N) = N, \qquad M = X = N; \tag{3}$$

 – for $N > 64$:

$$X(N) = \text{round}\,(59.249 + \sqrt{3510.39 - 9.50(431.14 - N)}\,),$$

$$M = \text{round}\,(0.10526X^2 - 12.473X + 431.14), \quad (X < 255), \tag{4}$$

$$M = 4085, \qquad\qquad\qquad\qquad\qquad\qquad (X = 255),$$

where N is the CCD camera output (Digital Number or DN), X is the 8-bit compressed number, M is the 12-bit decompressed value.

Ignoring read noise, dark current, and spurious charge, which are on the order of 1 DN, the data number is given by

$$N = \frac{nh\nu}{3.65c} + 11.5, \tag{5}$$

where n = number of photons detected, $h\nu$ = mean photon energy in eV, c = CCD camera gain constant of 100 electrons DN^{-1}; 11.5 is a digital offset in the CCD camera.

This yields the relation

$$n = \frac{(N - 11.5)\lambda}{34}, \tag{6}$$

and correspondingly

$$m = \frac{(M - 11.5)\lambda}{34}, \tag{7}$$

where λ is the photon wavelength in Å.

Finally, the compression error may be expressed in terms of photon statistics as

$$\varepsilon = \sqrt{\frac{\lambda}{34}} \, \frac{N - M}{\sqrt{N - 11.5}}. \tag{8}$$

3.1. IMAGE FORMATS

Images from the CCD camera of the SXT are transferred to image buffers in the SOLAR-A Data Processor (DP) for processing and read out to telemetry. There are three types of pictures, Full Frame Images (FFI), Partial Frame Images (PFI) and patrol images. In the following discussion a 'line of CCD data' will include 1024×1, 1024×2, or 1024×4 actual CCD pixels and be 1024, 512, or 256 output pixels long depending on whether 1×1, 2×2, or 4×4 on-chip summation is used. The FFI comprise 512, 256, 128, or 64 lines of CCD data so that at highest resolution (i.e., no on-chip pixel summation) they are, at most, only one half of a CCD picture. For on-chip summation modes, however, full CCD pictures can be taken. For PFI 64 lines of CCD data are always transferred to the buffer but, normally, only a subset of these data are sent to telemetry. In our terminology a PFI is a 64×64 sub-image that will include 64×64, 128×128, or 256×256 actual CCD pixels, depending upon the use of on-chip summation. For both FFI and PFI the band of CCD lines transferred to the DP is referred to as a Region of Interest (ROI). Patrol images are always 128 lines of CCD data acquired with 4×4 pixel summation (one-half of a CCD picture). Patrol images are used in the DP for Automatic Observing Region Selection (ARS). The patrol images are not normaly sent to telemetry.

For FFI the entire content of the image buffer is always sent to telemetry, i.e., is read out to the 10 Mbyte Bubble Data Recorder or transmitted directly to ground in real time. From the contents of the PFI buffer up to 16 adjacent (in the E–W direction) 64×64 pixel PFI are extracted from the 64 line image strips and sent to telemetry. Finally contiguous assemblies of PFI of any rectangular shape from a single PFI on up are called Observing Regions (OR). Four different shapes of observing regions are illustrated in Figure 12.

The arrangement and size of an X-ray or optical solar image on the CCD of the SXT

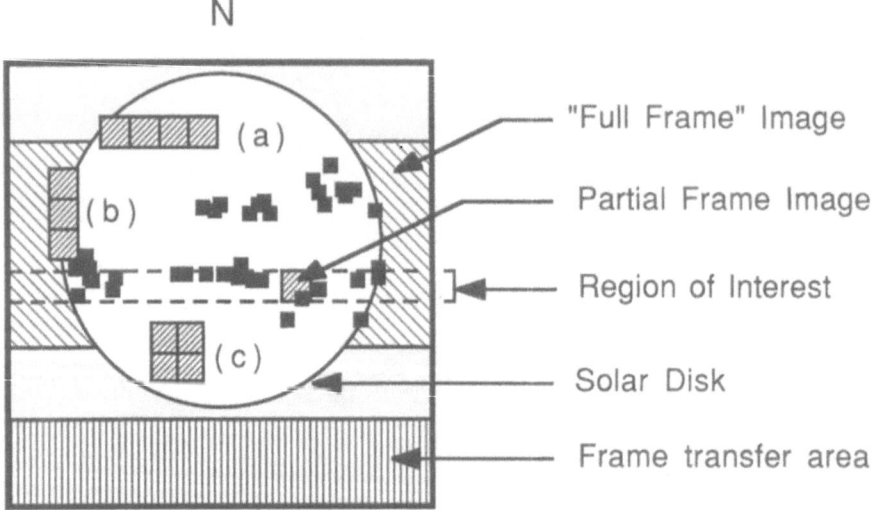

Fig. 12.   Arrangement of a solar image on the CCD. Ecliptic north is up and the solar west limb is to the left. Locations of 43 large flares observed by SMM in 1980 are indicated by black squares. The 'full frame' image comprises the largest number of pixels that can be sent to telemetry as an individual image. More typically, a partial frame image comprising 64×64 pixels, extracted from a designated region of interest, is sent to telemetry. In order to increase the size of an observing region the partial frame images may be grouped as illustrated by (a), (b), and (c). Observing regions of type (a) can be imaged with a single exposure whereas (c) requires 2 exposures and (b) requires 3 individual exposures. These examples are the size obtained when on-chip pixel summation is not used, i.e., at best angular resolution.

is shown in Figure 12. The vertical or column direction of the CCD is oriented towards ecliptic north. The solar image is deliberately de-centered upwards on the sensor to provide an unilluminated area at the bottom of the CCD. This area may serve as a frame transfer register to permit shutterless operation of the instrument in the event of trouble with the rotating mechanical shutter. In the frame transfer mode the portion of the picture to be saved would be rapidly shifted down into the frame transfer and then read into the image buffer at the normal rate of $131\,072$ pixels s^{-1}.

Considerable flexibility has been introduced into schemes of SXT image acquisition in order to make optimum use of the telemetry capacity of SOLAR-A. First, on-chip 2×2 and 4×4 pixel summation can be used to trade resolution for data rate. Second, the full frame images of 64, 128, 256, or 512 lines can be taken anywhere on the CCD to observe solar active longitudes of interest. Third, the region of interest of the partial frame images can be taken anywhere on the CCD and the 64×64 PFIs may be grouped together into larger Observing Regions (OR) as illustrated by examples (a), (b), and (c) of Figure 12. Observing regions of type (a) which are only extended E–W are acquired in a single exposure. Each block in the N–S direction, such as in examples (b) and (c), require a separate exposure. Although ORs with multiple-PFI extent N–S are assembled on the ground, they are treated as a single entity in planning and commanding. The angular extent of an FFI and a single PFI for the 3 summation modes are given in Table V.

3.2. EXPERIMENT CONTROL

SXT operation is under the control of software running in the SOLAR-A data processor (DP) computer and the SXT microprocessor. It is the function of the DP to command the SXT and to receive, format, and transfer all data to telemetry. It is the function of the SXT microprocessor to control the SXT mechanisms and CCD camera as directed by the DP. Command and status information is exchanged between the two computers via a block of memory in the SXT microprocessor which serves as a 'mailbox'.

The DP has the capability to examine SXT image data to select regions of interest and to adjust the length of the SXT exposure. It also monitors data from the hard X-ray, soft X-ray, or Bragg crystal spectrometers and may issue a flare flag to initiate SOLAR-A flare mode.

3.2.1. Sequence Tables

At the most fundamental level, the following parameters must be specified for each SXT exposure:
  - Location and size of portion of CCD image to be sent to telemetry;
  - on-chip CCD pixel summation of 1×1, 2×2, or 4×4;
  - exposure duration;
  - filter(s) to be used.

In addition, there are secondary instrument functions which may need to be altered and parameters that must be set (in all about 20 items).

The setup of the SXT for each observation is accomplished from tables of parameters stored in the DP called sequence tables. Each table provides all of the information necessary to define an SXT observing sequence, which can be quite complex. These tables, in turn, address other tables of fixed or adjustable control parameters. A sequence table is organized as a set of nested 'do-loops' in which a sequence of commands is executed repetitively until the prescribed number of cycles has been completed, at which time the program continues on to the next task.

This structure is illustrated in Table IV along with an example of an observing sequence for study of the evolution of a medium-sized active region. The sequence starts off with a single 10 × 10 arc min, 5 arc sec resolution, image centered on the observing region. This serves to place subsequent smaller images in a broader solar context. The large image is followed by a sequence of 5 × 5 arc min images at full resolution through each SXT filter. In this example there is no looping within the sequence. The sequence takes about 1.5 min per execution and is arranged to provide thin-filter images and optical images at uniform cadence at twice that rate. This enhances our ability to observe small white-light flares and low-level activity.

The image intervals, dt, include set-up and data transfer time as well as exposure time. Except for the thick beryllium filter the intervals are dominated by the time to transfer data to telemetry. The interval dt can either be held fixed, with the automatic exposure control permitted to operate only within this constraint, or allowed to increase in response to the automatic exposure software in the DP. Note that while this simple example concentrates on a single target, every numbered image in the sequence is independent and could image any one of the locations stored in the nine observing-region registers. Thus, it is feasible to continuously monitor the activity at several locations on the Sun in quiet mode.

There are 8 sequence tables stored in non-volatile memory in the DP, 4 each for partial frame (PFI) and full frame (FFI) images. At any given time 6 tables are designated for use under the 4 SOLAR-A science modes, QT/HIGH, QT/MED, FL/HIGH, and FL/MED. Sequence tables optimized for quiet-Sun or flare observations are sometimes quite different. Furthermore, the design of the cadence (number of loops) depends on the telemetry bit rate so the sequence table may be different between FL/HIGH mode (impulsive phase) and FL/MED mode (gradual phase) following the FL/HIGH mode. An 'entry table' is used to point to the desired sequence table for specific DP modes as shown in the following example:

Entry table

Data mode	TLM rate	Seq. table
QUIET (QT)	HIGH	FFI No. 1
QUIET (QT)	HIGH	PFI No. 1
QUIET (QT)	MED	FFI No. 2
QUIET (QT)	MED	PFI No. 2
FLARE (FL)	HIGH	PFI No. 3
FLARE (FL)	MED	PFI No. 4

For PFI observations in this example, No. 1 sequence table is used in the QT/HIGH mode. This will be automatically switched to sequence table No. 3 when FL/HIGH mode starts. (In this case, the sequence table No. 1 is optimized for QT/HIGH mode, and No. 3 for FL/HIGH mode.) If the bit rate is changed to MED in the gradual phase of the FL mode, the sequence table is switched to No. 4. The entry table numbers may be modified by real time or stored commands to achieve observing objectives. For instance, if we specify the sequence table No. 1 in the FL/HIGH column, then the same experiment sequence will continue regardless of the DP mode change to FL mode.

3.2.2. *Automatic Observing Region Selection*

The Automatic OR Selection (ARS) function utilizes patrol images taken for this purpose at regular intervals set by time commands or in response to a flare flag. There are two different ARS algorithms available when in the Quiet (QT) Mode and one algorithm while in the Flare (FL) Mode. Seach mode ARS is used to search the whole solar image for the four brightest sources in both QT and FL modes. Tracking mode ARS is used to track a feature as the region moves due to solar rotation, spacecraft pointing drift, or as the location of brightest emission evolves in QT mode. The QT ARS and FL ARS have separate parameters available to set up the exposure characteristics. The FL ARS has available its own automatic exposure control to help deal with the inordinate variation in X-ray brightness of flares. The accuracy of ARS source location is normally one patrol image pixel or 10 arc sec.

A patrol image is 42×21 arc min in size with a 10 arc sec (4×4 pixel) resolution. The area of the Sun covered in the search is therefore equivalent to the area labelled 'Full Frame Image' in Figure 12. As can be seen by comparison to the SMM flare locations plotted there, this adequately encompasses the active latitudes of the Sun.

The patrol image has highest priority and will interrupt acquisition of FFI or PFIs. The taking of patrol images is set in absolute time with an interval of 16 s to over an hour. Thus, by setting the time interval very long it is possible to prevent the taking of patrol images from interrupting, say, a movie sequence. QT and FL patrol images have independent cadences. It is possible to inhibit both patrol images and the ARS function.

There are nine observing region (OR) target registers in the DP to specify the locations of SXT observing regions. Five of these are filled by the ARS function and four by ground command. If ARS is inhibited, all nine registers may be loaded by ground command. Each register contains the location of the OR on the CCD and the number and shape of the mosaic of PFIs comprising an observing region.

The ninth OR register is redundantly filled with the location of the brightest feature by QT mode ARS to serve as the starting point for a flare search in response to a flare flag. The CCD exposure in process is aborted, a patrol image is taken and the FL ARS is activated at the time of the flare flag. The object is to center the flare observing region on the brightest feature as early as possible. If the flare is in the same observing region as was stored in the ninth OR register, flare observations will commence within 4 s. For an arbitray flare location the repositioning of the OR may require up to 6 s. The OR revised flare location is finally stored in the flare OR register.

Observing region (OR) table

OR No.	Location updated by
0	Quiet mode ARS
1	Quiet mode ARS
2	Quiet mode ARS
3	Quiet mode ARS
4	Quiet mode ARS
4	ART
5	ART
6	ART
7	ART
8	Flare mode ARS

3.2.3. *Automatic Observing Region Tracking*

While ARS will be useful to track bright X-ray structures on the Sun, sometimes it is necessary to keep the OR centered on a location, e.g., a coronal hole or emerging active region, that is not the brightest X-ray feature in the neighborhood. The Automatic OR Tracking (ART) software provides this function by tracking the movements of observing regions due to spacecraft attitude drift. Instead of changing the pointing of the telescope, the same function is served by choosing a diferent part of the CCD image. ART software uses the fine Sun sensor data of the SOLAR-A attitude control system for this purpose. At high data rate the correction is applied every 32 s. ART only works on the ground-commanded OR locations, not those selected by ARS – unless the contents of those registers are manually transferred to the four non-ARS registers.

3.2.4. *Automatic Exposure Control*

Because of the tremendous range in the X-ray brightness (e.g., Table III) of solar phenomena and the rapid change in X-ray intensity during flares it has been necessary to incorporate an Automatic Exposure Control (AEC) function in the SXT software. Every numbered image in a sequence table (Section 3.2.1) has its own AEC. This is implemented as follows. As a given image is read out from its image buffer to telemetry the number of pixels with intensity above an upper and the number below a lower threshold are counted and recorded. If the number of overbright pixels exceeds the table value (typically 10 pixels) the exposure is decreased or, if that is not possible, a thicker filter is selected. If the over-exposure test is passed, then the number of underbright pixels is compared to the table value (typically 100). If under-exposure is indicated, i.e., there are more pixels below the lower threshold than allowed, the exposure is increased or a thinner filter is selected for the next execution of that sequence table entry.

The AEC software runs in the DP which is a very busy real-time computer. However, the AEC algorithm is very fast so it takes only 6 s to update the exposure from the time the exposure was made. Since the AEC is a feed-back system with time delay, to assure stability of the AEC loop it is necessary that there be enough intervening images in the sequence table to allow exposure adjustment before the same sequence table entry is again commanded. If necessary, say for a high cadence movie in which only a single kind of picture was desired, this can be accomplished by simply duplicating the desired parameters in at most 3 sequence table positions and setting the rest of the table positions to 'No operation'.

The AEC only works on partial frame images (PFI) or observing regions comprised of PFIs. The actual implementation of the AEC is more complicated than this simple description implies. The logic tree for achieving proper exposure under the formidable range of observing conditions experienced by the SXT is quite complex.

3.3. IMAGE CADENCE

The time resolution of the SXT depends in a complex way upon many different factors, some determined by the experiment itself and some dictated by things such as the time since a downlink contact and the occurrence of flares. SOLAR-A science operation utilizes two SXT control and telemetry modes. In the quiet (QT) mode 62.5% of available telemetry is devoted to SXT images because the Hard X-ray Telescope does not produce science data (cf. Ogawara *et al.*, 1991). In flare (FL) mode, SXT uses 50% of the telemetry. In QT mode interleaved full frame and partial frame images of up to eight different observing regions may be acquired. In FL mode FFI are not taken. In either mode images are acquired at a cadence determined by the rate of transfer of the data to telemetry. SXT housekeeping and status data are telemetered regularly in addition to the image data. These status data contain image header information such as filter combination and exposure time which are telemetered in synchronism with the image data.

In normal QT mode both FFI and PFI are acquired, with periodic interruption by patrol images. The priority order for image acquisition is (1) patrol, (2) FFI, and (3) PFI. This normally results in gaps in PFI cadence as schematically illustrated in Figure 13. Note that a patrol image is automatically taken at the time of switch to FL mode. If there is a high scientific priority for obtaining PFI movies with no gaps, the control structure of SXT makes this possible at the sacrifice of FFI and the taking of patrol images. It is also possible to structure the SXT sequence control tables to continue the QT PFI cadence on into the FL mode in order to provide unchanged monitoring of a specific region.

A switch from QT to FL mode can occur because of a flare flag or by command. The transition from FL to QT mode normally occurs because energetic flare emission has dropped below preset thresholds and preset time intervals have expired. See Ogaware *et al.* (1991) for a description of use of the flare flag and the SOLAR-A bubble data recorder to store and sometimes overwrite less important data in response to flares.

3.3.1. *Quiet Mode*

In quiet (QT) mode SXT always acquires both full frame (FFI) and partial frame (PFI) images. The cadence of exposures is strictly determined by telemetry transfer time unless the exposure and set-up time exceed this, in which case a dummy image is transmitted.

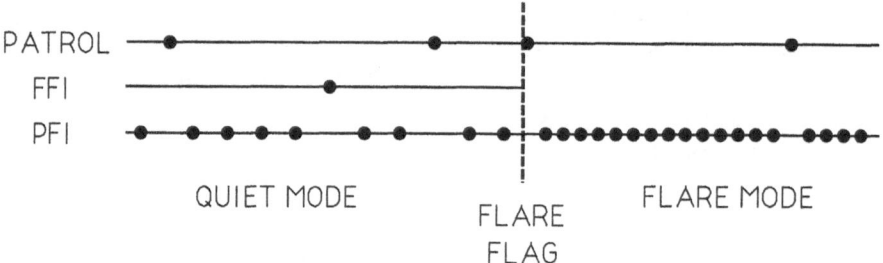

Fig. 13.   Scheme of interleaving the 3 different types of SXT exposures. Each filled circle represents a CCD exposure. Time runs from right to left. This is only a schematic example, the relative frequency of the different types of exposures is subject to control over a wide range.

There are two telemetry modes. In the PFI dominant mode, PFI data are transferred with telemetry rate 4 times faster than that of FFI, while in the FFI dominant mode, *vice versa*. This effect is evident in the times tabulated in Table V. The observing sequence example given in Table IV corresponds to the case of PFI dominant mode and telemetry rate high.

Separate image registers or buffers are used within the DP to receive pictures from the CCD camera and pass the desired portions on to telemetry. Although FFI and PFI may use either buffer A or B, it is not possible to combine the telemetry allocated to A and B to achieve yet higher time resolution, nor is it possible to devote both buffers simultaneously to only FFI or PFI.

For the fastest PFI cadence of 2 s image^{-1} the maximum allowable exposure duration is approximately 0.5 s in order to provide time to read out the CCD image and position the filter wheels. Table V provides a summary of information on SXT pixel size, image size, time resolution and maximum exposure times for basic full frame and partial frame images. Note that the very long maximum exposure times given for full frame images, while formally correct, are not generally desirable because of image smearing from pointing drift and waste of telemetry, i.e., no PFI can be taken during the long FFI exposure. Only very faint features such as the interior of coronal holes will require exposures greater than 10 or 20 s.

3.3.2. *Flare (FL) Mode*

In flare (FL) mode SOLAR-A assigns 12.5% of telemetry (which in QT mode is available to SXT) to the Hard X-ray Telescope and commands high telemetry rate for a predefined time interval. As illustrated in Figure 13, the taking of FFI is stopped, although the current contents of the FFI image buffer will eventually be transferred to telemetry and are not lost. In FL mode the interval between SXT images will be determined by the number of PFI in the observing region. For a single PFI the image cadence is 2 s at high telemetry rate and 16 s at medium rate.

TABLE IV

SXT sequence table structure

Basic sequence table

- LOOP 1 (n = ∞) — Image 1-1
- LOOP 2 (n = n2)
  - Image 2-1
  - Image 2-2
- LOOP 3 (n = n3)
  - Image 3-1
  - Image 3-2
  - Image 3-3
  - Image 3-4
- LOOP 4 (n = n4)
  - Image 4-1
  - Image 4-2
- LOOP 5 (n = n5)
  - Image 5-1
  - Image 5-2
  - Image 5-3
  - Image 5-4

PFI sequence table (example)

	Filter	Pixel sum	Obs. region	dr (s)
Image 1-1	Al 1265 Å	2 × 2	2 × 2	8
(n2 = 1)				
Image 2-1	4308 Å	1 × 1	2 × 2	8
Image 2-2	4600 Å	1 × 1	2 × 2	8
(n3 = 1)				
Image 3-1	Al/Mg/Mn	1 × 1	2 × 2	8
Image 3-2	Mg 2.52 μm	1 × 1	2 × 2	8
Image 3-3	Al 11.6 μm	1 × 1	2 × 2	8
Image 3-4	Al 1265 Å	1 × 1	2 × 2	8
(n4 = 1)				
Image 4-1	4308 Å	1 × 1	2 × 2	8
Image 4-2	4600 Å	1 × 1	2 × 2	8
(n5 = 1)				
Image 5-1	Be 119 μm	1 × 1	2 × 2	16
Image 5-2	Open	1 × 1	2 × 2	8
Image 5-3	NOP			0
Image 5-4	NOP			0
			Total sequence time = 96 s	

NOP = no operation = skip.

TABLE V

SXT image size, time resolution and exposure time

Telemetry		On-CCD sum	Pixel (arc sec)	Partial frame image		
Rate	Mode			Image size (arc min)	Time resol. (s)	Max. expos. (s)
High	PFI dom.	1 × 1	2.45	2.6 × 2.6	2.0	0.5
		2 × 2	4.91	5.2 × 5.2	2.0	0.5
		4 × 4	9.81	10.5 × 10.5	2.0	0.5
High	FFI dom.	1 × 1	2.45	2.6 × 2.6	8.0	6.5
		2 × 2	4.91	5.2 × 5.2	8.0	6.5
		4 × 4	9.81	10.5 × 10.5	8.0	6.5
Med	PFI dom.	1 × 1	2.45	2.6 × 2.6	16.0	14.5
		2 × 2	4.91	5.2 × 5.2	16.0	14.5
		4 × 4	9.81	10.5 × 10.5	16.0	14.5
Med	FFI dom.	1 × 1	2.45	2.6 × 2.6	64.0	62.5
		2 × 2	4.91	5.2 × 5.2	64.0	62.5
		4 × 4	9.81	10.5 × 10.5	64.0	62.5

Telemetry		On-CCD sum	Pixel (arc sec)	Full frame image		
Rate	Mode			Image size (arc min)	Time resol. (min)	Max. expos. (s)
High	FFI dom.	1 × 1	2.45	41.9 × 20.9	4.27	254.5
		2 × 2	4.91	41.9 × 41.9	2.13	126.5
		4 × 4	9.81	41.9 × 41.9	0.53	30.5
High	PFI dom.	1 × 1	2.45	41.9 × 20.9	17.07	1022.5
		2 × 2	4.91	41.9 × 41.9	8.53	510.5
		4 × 4	9.81	41.9 × 41.9	2.13	126.5
Med	FFI dom.	1 × 1	2.45	41.9 × 20.9	34.13	2046.5
		2 × 2	4.91	41.9 × 41.9	17.07	1022.5
		4 × 4	9.81	41.9 × 41.9	4.27	254.5
Med	PFI dom.	1 × 1	2.45	41.9 × 20.9	136.53	8190.5
		2 × 2	4.91	41.9 × 41.9	68.27	4094.5
		4 × 4	9.81	41.9 × 41.9	17.07	1022.5

3.3.3. *SXT High Time-Resolution Mode*

It is possible that high-resolution pictures could reveal quite fast flare phenomena in soft X-rays. To permit investigation of this possibility a special high time-resolution mode has been implemented within the SXT. In this case a PFI is divided horizontally (E–W) into 2 or 4 sub-images, one of which is acquired at twice or four times the normal rate. Each sub-image is a separate exposure timed, and with its region of interest (ROI) selected, by logic in the SOLAR-A microprocessor. As these 16 or 32 line ROIs are transferred to the DP they are combined into what the DP thinks is a normal 64-line

PFI. The disassembly of the images into a high time resolution (0.5 or 1 s) movie of 16 × 64 or 32 × 64 pixel images is done on the ground.

In order to maintain the fast cadence of this the high time-resolution mode certain restrictions apply. Exposures must be short and filter alternation is not possible from one frame to the next. Only a single observing region can be used but the usual on-chip pixel summation is permitted. Automatic exposure control is allowed under the same considerations as for a normal PFI.

4. Conclusion

The opportunity to take scientific instruments beyond the atmosphere of the Earth has enabled scientists to gain new knowledge on a grand and beautiful scale. The X-ray telescopes of the Skylab missions in the decade of the 1970s provided humankind's first extended look at phenomenon in the hot, dynamic outer atmosphere of a star. With the SOLAR-A soft X-ray telescope, thanks to improved technology, we have the opportunity to advance beyond Skylab with a simpler, smaller and less costly instrument. As we have described in the previous sections, the SXT provides innovative features catering to a wide range of observing targets. Thanks to phenomenal advances in data processing technology the SXT images should be much more easily and quickly accessible to the human mind than the earlier observations. For the best use of such elaborations, we hope that the SXT as well as the SOLAR-A mission survives for longer than is nominally expected, since it may be possible to study the solar-cycle dependency of various coronal parameters with a degree of reliability not previously attained. For flare studies the combination of SXT and the other SOLAR-A instruments provide a scientific capability much greater than the sum of the parts. We consider ourselves very fortunate to be able to contribute to these investigations and are eager to begin the mission.

Acknowledgements

We are grateful to our Japanese and American colleagues whose initiative and enthusiasm are responsible for the opportunity to have the SXT collaboration for the SOLAR-A mission. We especially recognize the leadership of Dr H. Hudson and the late Prof. K. Tanaka. We owe special thanks to our many scientific and engineering colleagues who have contributed to the design, fabrication, and testing of SXT. Among them K. Nariai and T. Watanabe for X-ray optics; T. Cruz and W. Rosenberg for aspect optics; K. Appert and D. Kyrie for electronics; D. Akin for shutter design; D. Murray, B. Costanzo, T. Hasui, and A. Hagiwara for thermal design; J. Vieira and K. Gowen for mechanical design; R. Stern for the CCD; C. Feinstein, W. Brooker, and L. Shing for test and calibration, I. Kondo for control system architecture; T. Kato and A. Yamaguchi for supporting SXT testing at ISAS; D. T. Roethig for computer and e-mail support; B. Rix, M. Finch, F. Friedlaender, D. Kauffman, and W. Jaynes for contract and program management; R. Fielder, S. Taylor, and T. Iwata for secretarial support, and N. Nitta for quick-look software.

We acknowledge the enhusiastic contribution in building the SXT portion of the SOLAR-A data processor (DP) by the Fujitsu DP/SXT team (I. Odaki, T. Tanaka, M. Kurihara, K. Nomura, and M. Kojima) led by S. Kubo. M. Abe of Systems Engineering Consultants Co., Ltd., led by I. Akiyama, made major contributions to software development. We also thank NEC Corp., especially M. Matsui and Y. Masumoto, for the integration of SXT to the SOLAR-A spacecraft.

The SXT program in the U.S. has been wisely and collegially managed and administered by C. Pellerin and J. Lintott of NASA headquarters and R. Ise and H. Hill of NASA Marshall Space Flight Center. The SXT CCD camera was provided by the Jet Propulsion Laboratory and we owe special thanks to L. Hovland, T. Bursch, J. Janesick, J. Daniels, M. Schwockert, T. Radey, E. Villegas, A. Collins, and R. White. It is only thanks to the expertise and cooperation of J. Hynecek and I. Fujii of Texas instruments that SXT has an excellent CCD. We have enjoyed outstanding teamwork and technical performance from our SXT subcontractors:

- *X-ray mirror*: United Technology Optical Systems (A. Slomba, R. Kusha, G. Hull-Allen, M. Laughlin).
- *Carbon fiber metering tube*: Fiber Technology Corp. (B. Lundy).
- *Filter wheel and shutter motor*: Shaeffer Magnetics.
- *X-ray filters*: Luxel Corp. (G. Steele, F. Powell).
- *Optical filters*: Andover Corp., Perkin–Elmer.

The program has been supported at Lockheed under contract NAS8–37334 with NASA Marshall Space Flight Center and by the Lockheed Independent Research Program.

References

Acton, L., Morrison, M., Janesick, J., and Elliott, T.: 1991, *Proc. SPIE, Charge-Coupled Devices and Solid State Optical Sensors* **1447**, 123.
Bendinelli, O.: 1991, *Astrophys. J.* **366**, 599.
Bruner, M. E., Acton, L. W., Brown, W. A., Stern, R. A., Hirayama, Y., Tsuneta, S., Watanabe, T., and Ogawara, Y.: 1989, in J. H. Waite, J. L. Birch, and R. L. Moore (eds.), *Proceedings of the 1988 Yosemite Conference on Outstanding Problems in Solar System Plasma Physics: Theory and Instrumentation*, Am. Geophysical Union Monograph, Vol. 54, p. 187.
Craig, I. J. D. and Brown, J. C.: 1986, *Inverse Problems in Astronomy*, Adam Hilger Ltd., Bristol and Boston.
Fludra, A. and Sylwester, J.: 1986, *Solar Phys.* **105**, 323.
Hynecek, J.: 1979, *IEEE IEDM Tech. Dig.* **611**.
Janesick, J., Hynecek, J., and Blouke, M.: 1981, *Proc. SPIE, Solid-State Imagery for Astronomy* **290**, 165.
Janesick, J., Klaasen, K., and Elliott, T.: 1987, *Opt. Eng.* **26**, 972.
Kane, S. R., Anderson, K. A., Fenimore, E. E., Klebesadel, R. W., and Laros, J. G.: 1979, *Astrophys. J.* **233**, L151.
Lemen, J. R., Claflin, E. S., Brown, W. A., Bruner, M. E., Catura, R. C., and Morrison, M. D.: 1989, *Proc. SPIE, X-Ray/EUV Optics for Astronomy and Microscopy* **1160**, 316.
Lemen, J. R., Acton, L. W., Brown, W. A., Bruner, M. E., Catura, R. C., Strong, K. T., and Watanabe, T.: 1991, *Adv. Space Res.* (to be published).
McTiernen, J.: 1991, in R. Canfield and Y. Uchida (eds.), *Proceedings of K. Tanaka Memorial Symposium*, Springer-Verlag, Berlin (to be published).
Mewe, R., Gronenschild, E. H. B. M., and van den Oord, G. H. J.: 1985, *Astron. Astrophys. Suppl. Series* **62**, 197.

Mewe, R., Lemen, J. R., and van den Oord, G. H. J.: 1986, *Astron. Astrophys. Suppl. Series* **63**, 511.

Morrison, M. D., Lemen, J. R., Acton, L. W., Bentley, R. D., Kosugi, T., Tsuneta, S., Ogawara, Y., and Watanabe, T.: 1991, *Solar Phys.* **136**, 105 (this issue).

Nariai, K.: 1987, *Appl. Optics* **26**, 4428.

Nariai, K.: 1988, *Appl. Optics* **27**, 345.

Ogawara, Y., Takano, T., Kato, T., Kosugi, T., Tsuneta, S., Watanabe, T., Kondo, I., and Uchida, Y.: 1991, *Solar Phys.* **136**, 1 (this issue).

Sakurai, T.: 1990, in Y. Osaki and H. Shibahashi (eds.), *Progress of Seismology of the Sun and Stars*, Springer Lecture Notes in Physics, No. 367, p. 253.

Strong, K. T. and Lemen, J. T.: 1987, unpublished work.

Strong, K. T., Acton, L. W., Brown, W. A., Claflin, S. L., Lemen, J. R., and Tsuneta, S.: 1991, *Adv. Space Res.* (to be published).

Vaiana, G. S., Van Speybroek, L., Zombeck, M. V., Krieger, A. S., Silk, J. K., and Timothy, A.: 1977, *Space Sci. Instr.* **3**, 19.

Watanabe, T.: 1987, *Bull Tokyo Astron. Obs. 2nd Ser.* **277**, 3213.

THE WIDE BAND SPECTROMETER ON THE SOLAR-A*

M. YOSHIMORI, K. OKUDAIRA, Y. HIRASIMA, T. IGARASHI,
M. AKASAKA,
Y. TAKAI, K. MORIMOTO

Rikkyo University, Toshima-ku, Tokyo 171, Japan

T. WATANABE, K. OHKI

National Astronomical Observatory, Mitaka, Tokyo 181, Japan

J. NISHIMURA, T. YAMAGAMI, Y. OGAWARA

Institute of Space and Astronautical Science, Sagamihara, Kanagawa 229, Japan

and

I. KONDO

Shibaura Institute of Technology, Ohmiya, Saitama 330, Japan

(Received 22 March, 1991; in revised form 17 May, 1991)

Abstract. The SOLAR-A spacecraft has spectroscopic capabilities in a wide energy band from soft X-rays to gamma-rays. The Wide Band Spectrometer (WBS), consisting of three kinds of spectrometers, soft X-ray spectrometer (SXS), hard X-ray spectrometer (HXS) and gamma-ray spectrometer (GRS), is installed on SOLAR-A to investigate plasma heating, high-energy particle acceleration, and interaction processes. SXS has two proportional counters and each counter provides 128-channel pulse height data in the 2–30 keV range every 2 s and 2-channel pulse count data every 0.25 s. HXS has a NaI scintillation detector and provides 32-channel pulse height data in the 20–400 keV range every 1 s and 2-channel pulse count data every 0.125 s. GRS has two identical BGO scintillation detectors and each detector provides 128-channel pulse height data in the 0.2–10 MeV range every 4 s and 4-channel pulse count data (0.2–0.7, 0.7–4, 4–7, and 7–10 MeV) every 0.25–0.5 s. In addition, each of the BGO scintillation detectors provides 16-channel pulse height data in the 8–100 MeV range every 4 s and 2-channel pulse count data (8–30 and 30–100 MeV) every 0.5 s. The SXS observations enable one to study the thermal evolution of flare plasma by obtaining time series of electron temperatures and emission measures of hot plasma; the HXS observations enable one to study the electron acceleration and heating mechanisms by obtaining time series of the electron spectrum; and the GRS observations enable one to study the high-energy electron and ion acceleration and interaction processes by obtaining time series of electron and ion spectra.

1. Scientific Objectives

The SMM and HINOTORI satellites observed a large number of flares in the 21st solar maximum in the early 1980s. New findings of magnetic energy release processes, plasma heating, and particle acceleration were reported. The detailed scientific results were presented in the following special issues: *Solar Phys.* **86** (1983), No. 1, *Solar Phys.* **111** (1987), No. 1, *Solar Phys.* **113** (1987), Nos. 1 and 2, *Solar Phys.* **118** (1988), Nos 1 and 2, *Space Sci. Rev.* **51** (1989), No. 1, and *Astrophys. J. Suppl.* **73** (1990). Some of the primary questions concerning these fundamentals of flare phenomena, however, still

---

* After the launch the name of SOLAR-A has been changed to YOHKOH.

Solar Physics **136**: 69–88, 1991.
© 1991 *Kluwer Academic Publishers.*

remain unsolved. Attempts have been made to show how future observations with increased sensitivity and improved spatial and spectral resolutions will answer these questions. Explosive phenomena such as the solar flare are a common characteristic of cosmic plasma at many sites throughout the universe, and a detailed understanding of these high-energy processes is one of the major goals of astrophysics.

Spectroscopic observations in a wide energy band of soft X-rays to gamma-rays provide important clues to the plasma heating and particle acceleration processes. The soft X-ray spectrum consists of components of continuum and various lines. The emission in energies above 5 keV is mostly continuum which results from hot thermal plasma, with the exception of the strong line emission at 6.7 keV which results from excited iron ions. The electron temperature and emission measure of hot plasma can be derived from the soft X-ray continuum spectrum (e.g., Watanabe *et al.*, 1983). Further, the thermal energy content is calculated from these two parameters. The thermal evolution of the flare plasma can be studied with these temporally resolved quantities. The soft X-ray spectral observation diagnoses the hot flare plasma and provides complementary data for other experiments on board SOLAR-A: the soft X-ray telescope (SXT) and the Bragg crystal spectrometer (BCS).

The hard X-ray observation provides the most direct information for understanding the energy release processes and the electron acceleration mechanisms. The hard X-ray intensity variations correlate with electron acceleration. The onset of the hard X-ray emission gives the most important signal for the earliest time of the magnetic energy release. The electron acceleration and plasma heating mechanisms can be deduced from the temporal and spectral analyses of the hard X-ray emission (e.g., Dennis, 1988). The conditions at the energy release site, the magnetic field configuration, and the parameters of the flare plasma before, during, and after the flare are determined from coincident observations of images in both hard and soft X-rays and at other wavelengths.

The gamma-ray observations provide irreplaceable information for study of the highest-energy flare phenomena (Chupp, 1984, 1987; Yoshimori, 1989). The radiation in the gamma-ray range is emitted by several processes including high-energy electron bremsstrahlung, nuclear de-excitation, positron annihilation, neutron capture, and neutral pion decay. The electron bremsstrahlung produces a continuum spectrum extending to the maximum energy of the accelerated electrons. Thus the electron acceleration to the highest energies can be studied from the temporal and spectral analyses of the gamma-ray emission. The other gamma-ray emissions result from energetic ion interactions with the solar ambient medium. The detection of these gamma-ray emissions gives evidence of nuclear reactions at the flare site. The nuclear de-excitation produces a prompt line emission in the MeV energies. The nuclear de-excitation lines are mostly produced by ions accelerated to 10–30 MeV nucl^{-1}. The positrons are mostly emitted by beta-decay nuclei produced by 10–100 MeV nucl^{-1} ion interactions. The neutron capture on protons emits a delayed line at 2.22 MeV. The neutrons contributing to the 2.22 MeV line emission are produced by ions with energies <100 MeV nucl^{-1}. The pion decay gamma-ray spectrum with a broad peak at 70 MeV is different from the high-energy electron bremsstrahlung spectrum. Since the pions are produced by protons accelerated to energies > 1 GeV, the pion decay gamma-ray emission provides evidence of very high-energy proton acceleration. High-energy neutron observation also provides similar evidence. The high-energy electron and ion acceleration processes are investigated from the temporal and spectral analyses of the gamma-ray emission over wide energy ranges. Sufficiently detailed information cannot be determined from the gamma-ray observations alone and coincident observations of solar energetic particles (SEP) in interplanetary space and high-energy neutron measurements at ground-based stations are required to complement the data.

The Wide Band Spectrometer (WBS), which consists of three kinds of spectrometers covering the energy range from soft X-rays to gamma-rays, is installed on the SOLAR-A to observe the detailed spectrum and temporal evolution of this wide-band photon emission. It will help to better understand the important processes of plasma heating and particle acceleration in solar flares.

2. Instrument Description

The WBS consists of the following subinstruments: (1) soft X-ray spectrometer (SXS), (2) hard X-ray spectrometer (HXS), (3) gamma-ray spectrometer (GRS), and (4) radiation belt monitor (RBM). SXS, HXS, and GRS are solar instruments pointed at the Sun. RBM has the capability of detecting the radiation belt passage and is pointed perpendicular to the solar direction. Descriptions of the WBS instrument can be found elsewhere (Yoshimori, 1988; Yoshimori et al., 1988; Kondo et al., 1990).

2.1. SOFT X-RAY SPECTROMETER (SXS)

SXS consists of a gas proportional counter filled with Xe and CO_2 (1.16 atm) and covers the energy range of 2 to 30 keV. The gas proportional counter has three anode wires: two of them are connected together to form SXS-1 output and the third anode wire forms SXS-2 output. The cross-sectional view of the SXS detector is shown in Figure 1. The three anode wires are biased with one high-voltage power supply unit. SXS has a field of view of 10 deg × 10 deg (20 times the diameter of the Sun), reducing the background by an aluminum slit collimator. The SXS counters have beryllium windows of 150 μm thickness. The SXS-1 counter has an area of 11.88 cm^2 with the additional 150 μm thick aluminum filter and 0.0707 cm^2 without filters, while the SXS-2 counter has aluminum filters of 50 μm thickness (0.0353 cm^2) and 300 μm thickness (6.12 cm^2). The effective areas of SXS-1 and SXS-2 counters are calculated as a function of energy and shown in Figure 2. The SXS-1 counter has a large effective area suitable for detection of a small flare, whereas the SXS-2 counter has a small effective area suitable for detection of large flares. Thus the SXS, with the two different gas proportional counters, has the capability of detecting flares of various sizes.

The SXS spectral response to a 5.9 keV Mn Kα line from the [55]Fe radioactive source is shown in Figure 3. The energy resolution (FWHM) at 5.9 keV is about 20% and is shown as a function of energy in Figure 4. The in-flight energy calibration is achieved by detection of the 5.9 keV Mn Kα line of the attached [55]Fe calibration source.

Soft X-Ray Spectrometer

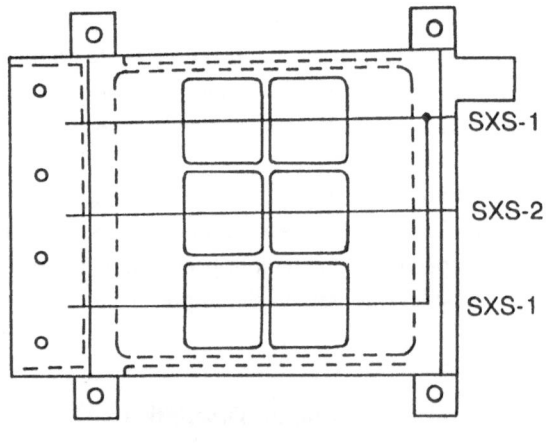

(a) top view

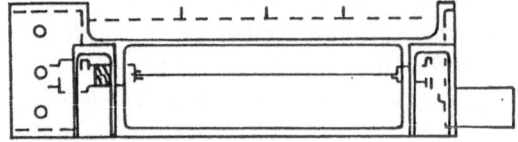

(b) bottom view

Fig. 1.   Cross-sectional view of SXS gas proportional counter. (a) Top view and (b) side view. The SXS gas proportional counter has three anode wires: two of them are connected together to form SXS-1 output and the third wire forms SXS-2 output.

The SXS electronic block diagram is shown in Figure 5. The SXS primary output data from both SXS-1 and SXS-2 counters are 128-channel pulse height data in the 2–30 keV range every 2 s (SXS-PH1 for SXS-1 and SXS-PH2 for SXS-2) and 2-channel pulse count data every 0.25 s (SXS-PC11 and 12 for SXS-1 and SXS-PC21 and 22 for SXS-2). The energy ranges of the pulse count data are changeable by command. One of the 4 pulse count data is also used to monitor the occurrence of solar flares.

2.2. HARD X-RAY SPECTROMETER (HXS)

The HXS consists of a NaI scintillator, which is 7.6 cm in diameter and 2.5 cm in thickness, optically coupled to a 7.6 cm diameter photomultiplier tube. The cross-sectional view of the HXS is shown in Figure 6. The HXS covers the energy range of

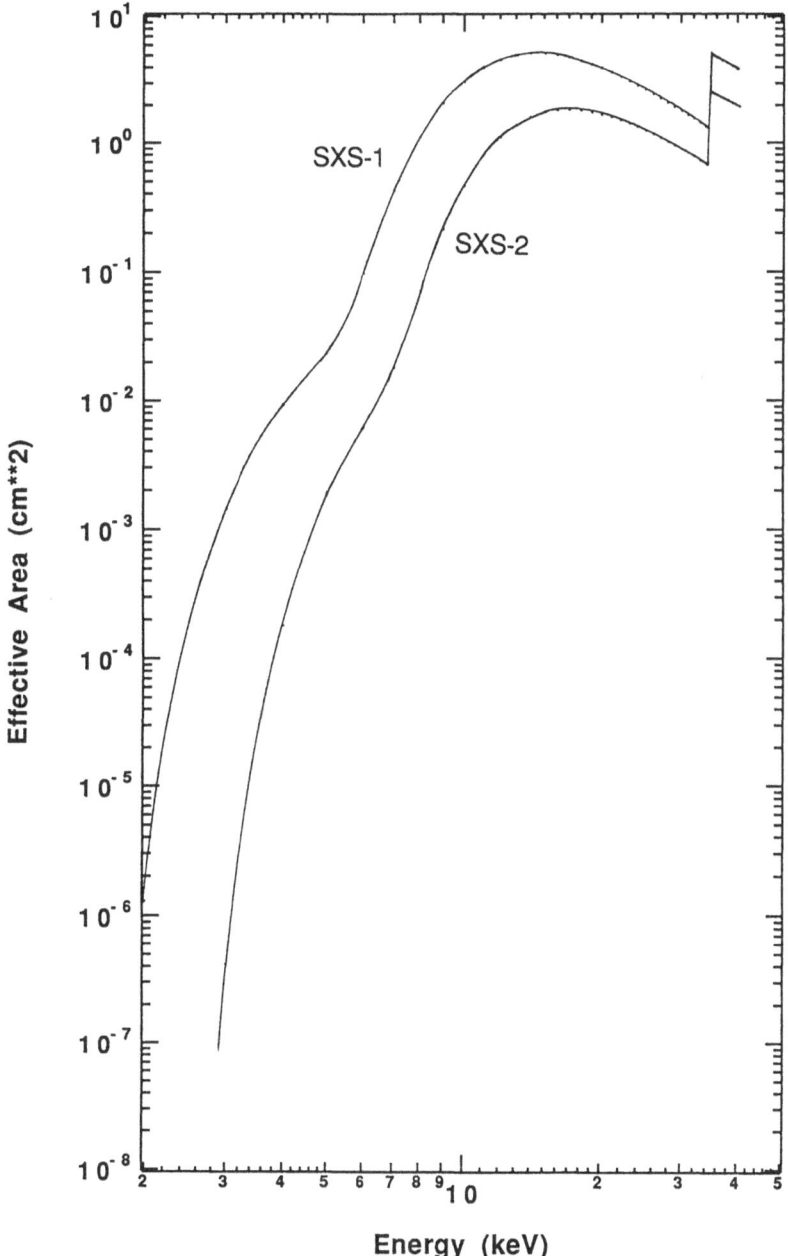

Fig. 2.   Calculated effective areas of SXS-1 and SXS-2 counters as functions of X-ray energy.

20–600 keV. The NaI scintillator is covered with two stainless steel absorbers (13.8 cm² × 0.08 mm thick + 31.8 cm² × 1 mm thick) to suppress low-energy X-ray events. The effective area of HXS is shown as a function of energy in Figure 7. The HXS spectral response to 30, 81, and 356 keV lines from ^{133}Ba electron capture is shown in

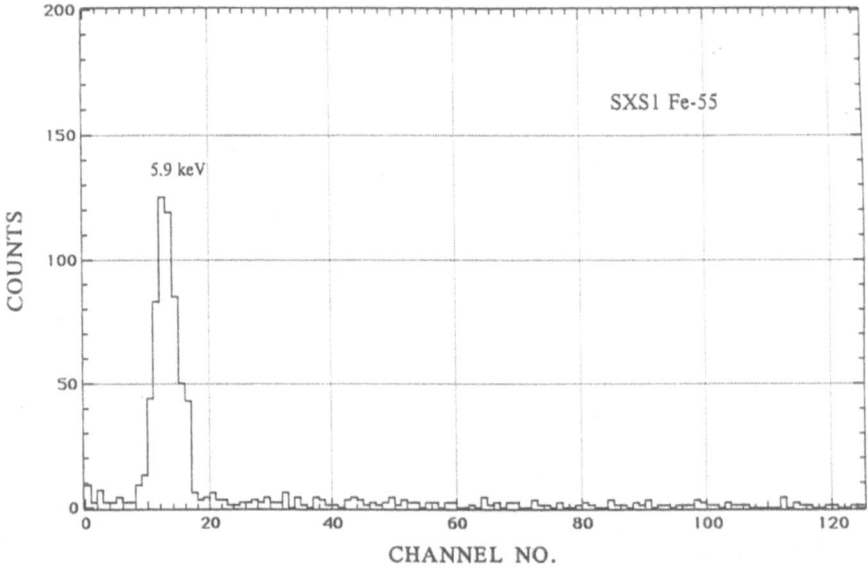

Fig. 3.   SXS spectral response to 5.9 keV Mn Kα line from ⁵⁵Fe radioactive source.

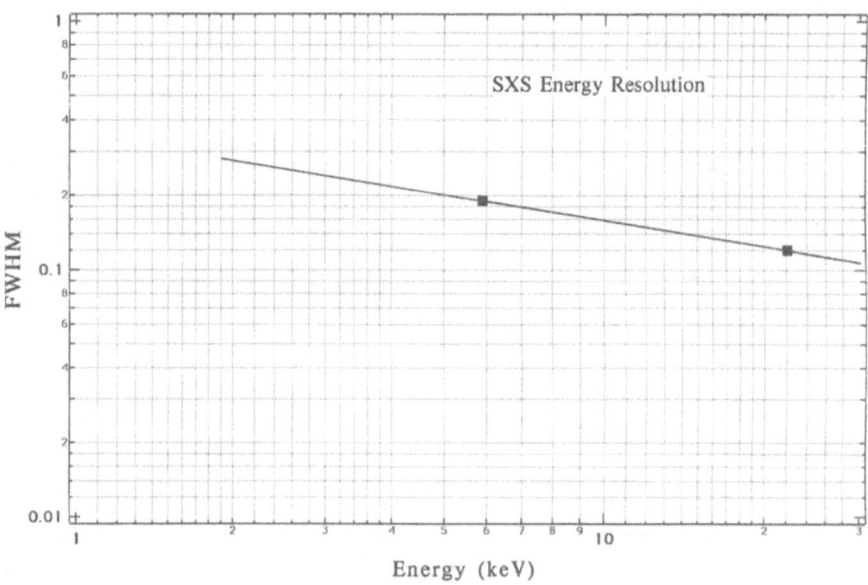

Fig. 4.   Energy resolution (FWHM) of the SXS counter as function of energy.

Figure 8. The energy resolution (FWHM) at 60 keV is 15% and is shown as a function of energy in Figure 9. The incident hard X-ray spectrum is obtained by deconvoluting the observed 32-channel energy-loss spectrum (Forrest, 1990). An HXS response matrix obtained from the Monte Carlo simulation is used for the spectral deconvolution. As an example, the incident hard X-ray spectrum of E^{-3} below 100 keV and E^{-5} above

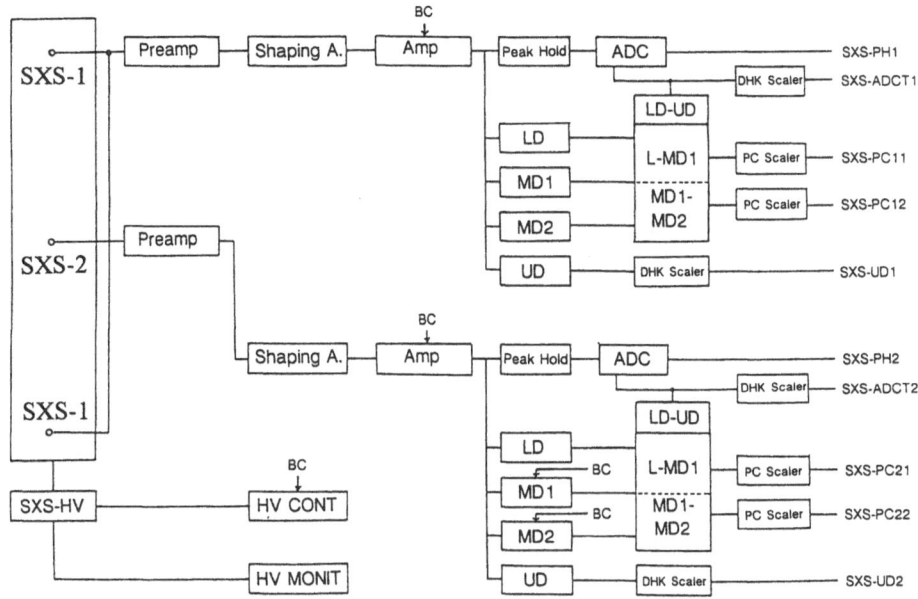

Fig. 5. SXS electronic block diagram. Three anode wires are biased by one high-voltage power supply unit (HV). Each of the SXS-1 and SXS-2 outputs is fed to a similar electronic circuit. Each of the SXS counter outputs passes through a preamplifier and shaping amplifier, and is fed to a main amplifier. After amplification, the output is sent to an 8-bit ADC to produce 128-channel pulse height data (SXS-PH). The amplified output is also sent to a discriminator with 4 levels (LD, MD1, MD2, and UD) to produce 2-channel pulse height count data (SXS-PC11 and 12 for SXS-1 and SXS-PC21 and 22 for SXS-2). In addition, ADC triggering count (SXS-ADCT), and pulse count above upper discriminator level (SXS-UD) are monitored as housekeeping data. HV level, main amplifier gain, and MD1 and MD2 discrimination levels are changeable by block command (BC).

100 keV and the convoluted energy-loss spectrum are shown in Figure 10. The in-flight energy calibration is achieved by detection of a 60 keV line from ^{237}Np nuclear de-excitation (^{237}Np is an ^{241}Am α-decay product). In this HXS calibration, the 60 keV line event in coincidence with an α-particle (5.48 MeV) event detected with two silicon detectors is recorded. The in-flight energy calibration method is schematically shown in Figure 11. The HXS in-flight energy calibration spectrum measured with the coincidence method is shown in Figure 12. The details of the in-flight energy calibration system was described earlier (Yoshimori and Okudaira, 1988).

The HXS electronic block diagram is shown in Figure 13. The HXS primary output is 32-channel quadratic-spaced pulse height data in the 20–400 keV range every 1 s (HXS-PH), and 2-channel pulse count data (HXS-PC1 (20–50 keV) and HXS-PC2 (50–600 keV)) every 0.125 s. HXS-PC1 is also used to monitor the occurrence of solar flares.

2.3. GAMMA-RAY SPECTROMETER (GRS)

The GRS consists of two identical BGO scintillators. Each scintillator is 7.6 cm in diameter and 5.1 cm in thickness and is optically coupled to a 7.6 cm diameter photo-

Hard X-Ray Spectrometer

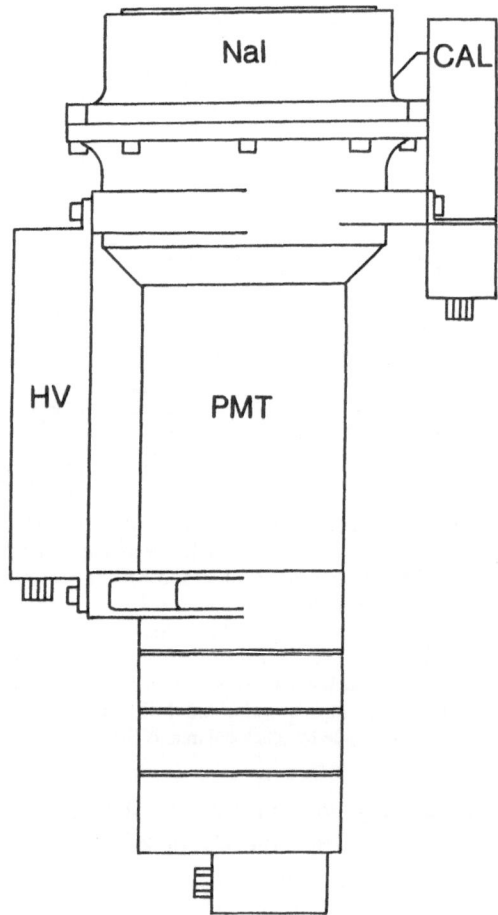

Fig. 6.   Cross-sectional view of HXS NaI detector. A NaI(Tl) scintillator of 7.6 cm diameter and 2.5 cm thickness is optically coupled with a photomultiplier tube (PMT) biased by a high-voltage power supply unit (HV). CAL is an in-flight energy calibration device.

multiplier tube. The cross-sectional view of GRS is shown in Figure 14. Since the BGO ($Bi_4Ge_3O_{12}$) scintillator has high density and high effective atomic number, these properties contribute toward improving the detection sensitivity to gamma-rays. Each BGO scintillator is covered with a 0.5 mm thick lead absorber to suppress low-energy gamma-ray events. The GRS spectral responses to 3.08 and 4.07 MeV lines from ^{49}Ca nuclear de-excitation (^{49}Ca nuclear excitation state is produced by thermal neutron-capture reaction of ^{48}Ca nuclei) and 12.79 and 17.23 MeV lines from ^{7}Be nuclear de-excitation (^{7}Be nuclear excited state is produced by nuclear reaction of $p(^7Li, ^7Be)n$)

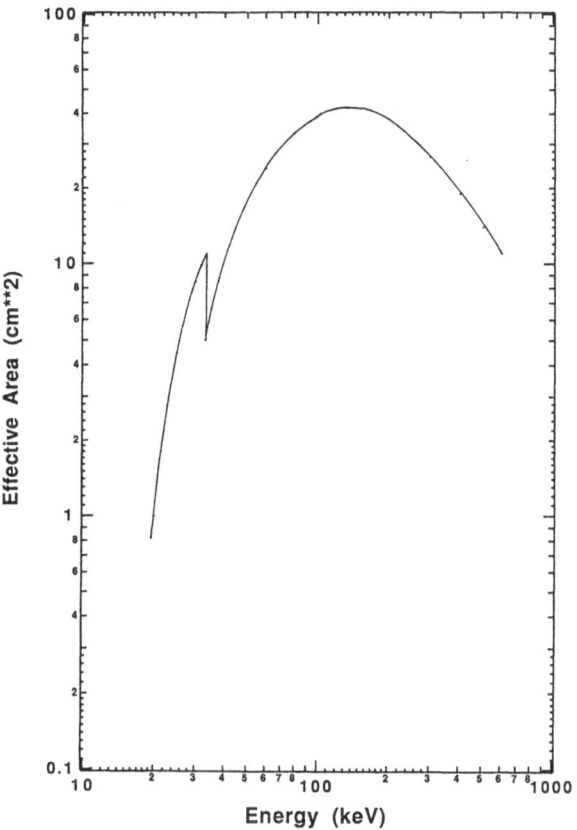

Fig. 7.   Calculated effective area of HXS as a function of X-ray energy.

are shown in Figures 15(a) and 15(b). The energy resolution (FWHM) of GRS is shown as a function of energy in Figure 16. The BGO energy resolution is about twice that of the NaI scintillator due to the low scintillation efficiency of the BGO scintillator. The energy dependence of the effective area of GRS is calculated as a function of gamma-ray energy by the Monte-Carlo simulation. The result is shown in Figure 17. The incident gamma-ray spectrum is obtained by deconvoluting the observed 128-channel energy-loss spectrum. The GRS response matrix obtained from the Monte Carlo simulation is used for the spectral deconvolution. As an example, the incident gamma-ray spectrum of E^{-3} continuum + lines and the convoluted energy-loss spectrum are shown in Figure 18.

The GRS electronic block diagram is shown in Figure 19. Each of two identical GRS detectors (GRS-1 and GRS-2) is connected to a similar electronic circuit. Since GRS covers the wide energy range of 0.2–100 MeV, the 0.2–10 MeV output pulse (GRS-L) and 8–100 MeV output pulse (GRS-H) are produced from the anode and the 6th dynode of the photomultiplier tube, respectively. The primary output data of GRS-L are 128-channel quadratic-spaced pulse height data every 4 s (GRS-PHL1 and

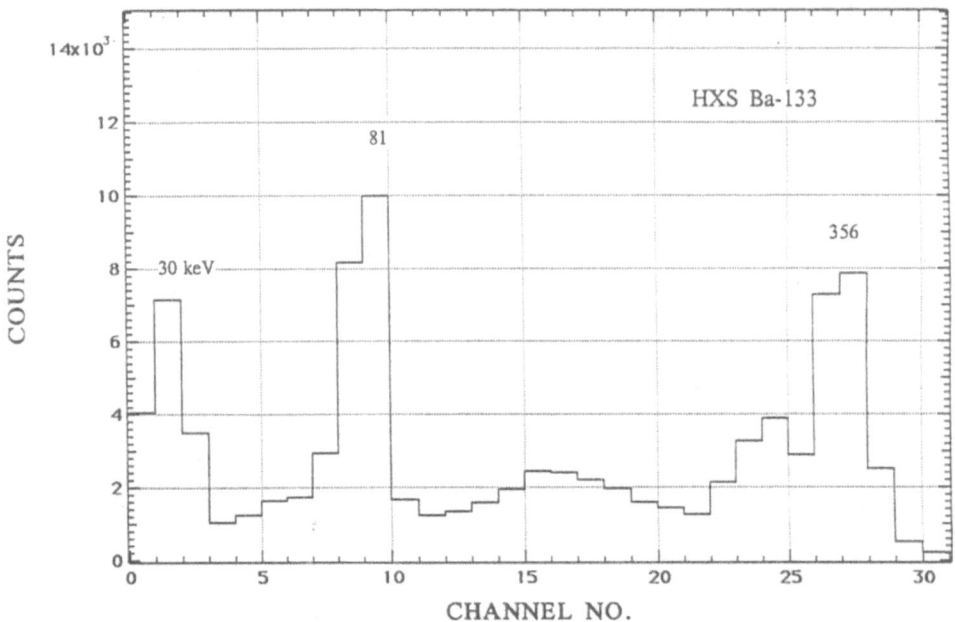

Fig. 8.   HXS spectral response to 30, 81, and 356 keV lines from ^{133}Ba electron capture.

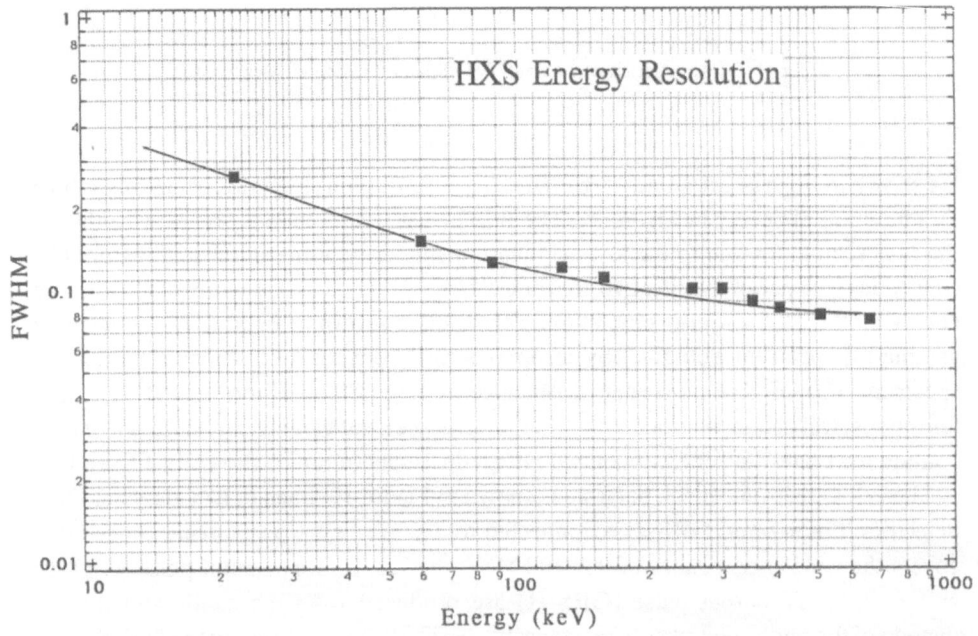

Fig. 9.   Energy resolution (FWHM) of HXS as a function of energy.

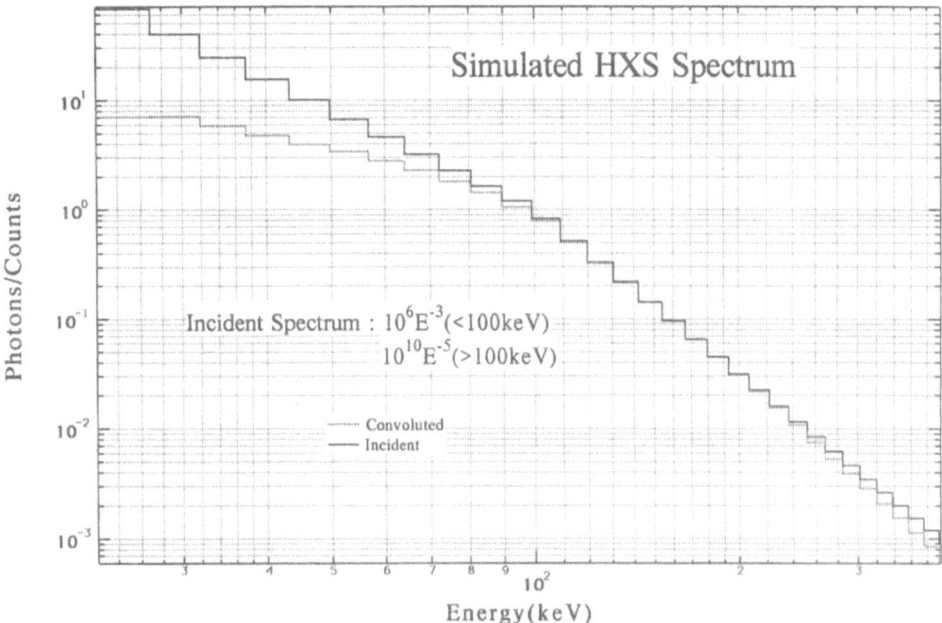

Fig. 10. Hard X-ray incident spectrum (E^{-3} below 100 keV and E^{-5} above 100 keV) and convoluted energy-loss spectrum.

GRS-PHL2) and 4-channel pulse count data (GRS-PC11 and 21 for 0.2–0.7 MeV, GRS-PC12 and 22 for 0.7–4 MeV, GRS-PC13 and 23 for 4–7 MeV and GRS-PC14 and 24 for 7–10 MeV). The time resolution is 0.25 s for GRS-PC11, 12, 21, and 22, and 0.5 s for GRS-PC13, 14, 24, and 25. The primary output data of GRS-H are 16-channel

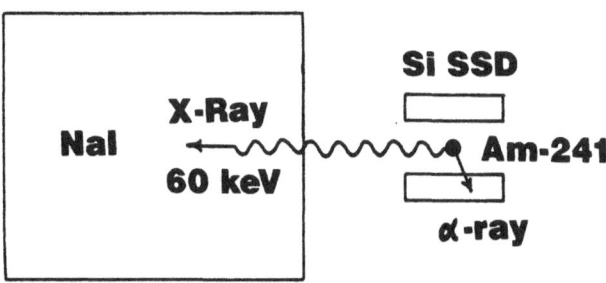

Fig. 11. HXS in-flight energy calibration method is schematically shown. 60 keV nuclear de-excitation line and 5.48 MeV α-ray are simultaneously emitted from ^{241}Am radioactive source. 60 keV line detected with HXS NaI(T1) scintillator in coincidence with 5.48 MeV α-ray detected with two silicon solid state detetors (Si SSD) is recorded as calibration data.

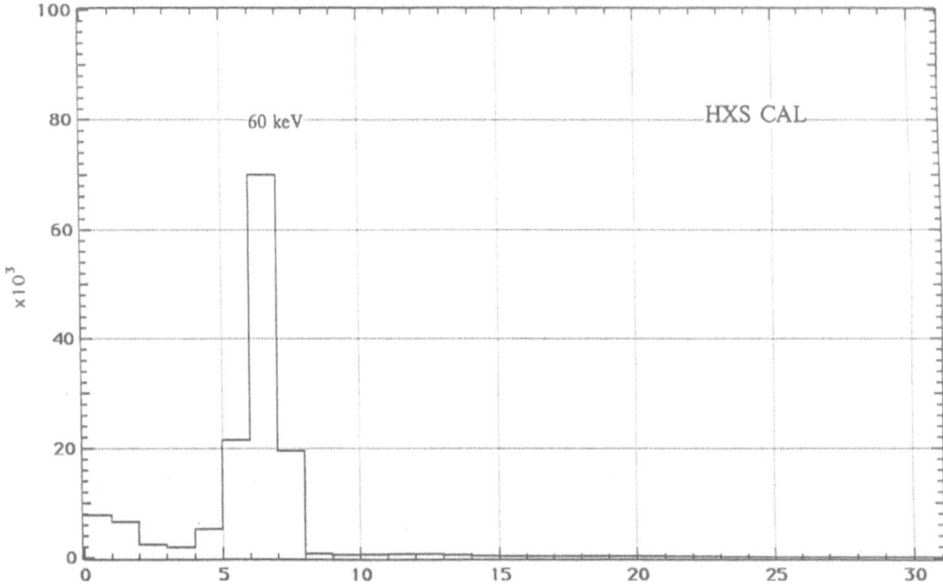

Fig. 12. HXS in-flight energy clibration spectrum of 60 keV line from ^{237}Np nuclear de-excitation.

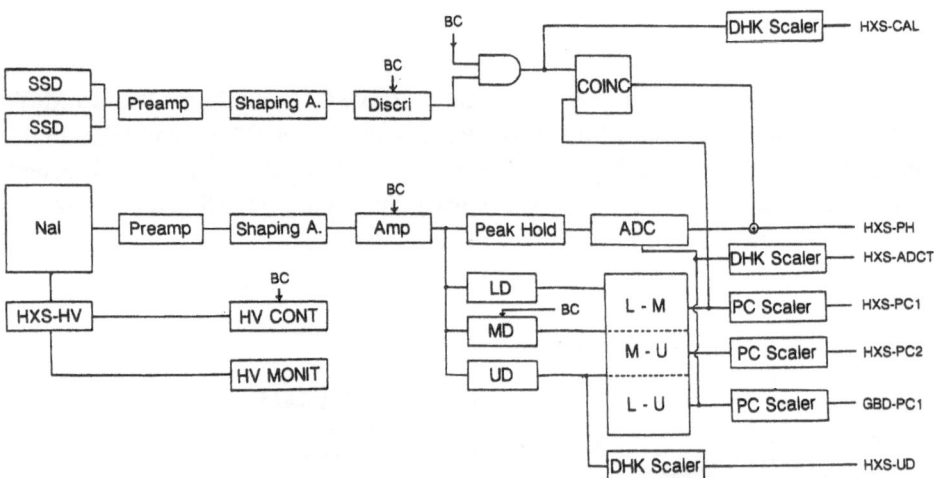

Fig. 13. HXS electronic block diagram. NaI(T1) scintillation detector output passes through a pre-amplifier and shaping amplifier, and is fed to a main amplifier. After amplification, the output is sent to an 8-bit ADC to produce 32-channel pulse height data (HXS-PH). The amplified output is also sent to a discriminator with 3 levels (LD, MD, and UD) to produce 2-channel pulse count data (HXS-PC1 and HXS-PC2). Silicon solid state detector (SSD) output passes through a preamplifier and shaping amplifier, and is fed to a discriminator. A coincidence event between NaI(T1) and SSD detectors provides an in-flight energy calibration event. In addition, ADC triggering count (HXS-ADCT), pulse count above upper discrimination level (HXS-UD), and SSD pulse count (HXS-SSD) are monitored as house-keeping data. HV level, main amplifier gain, SSD discrimination level, and NaI(T1) discrimination level are changeable by block command (BC).

Gamma-Ray Spectrometer

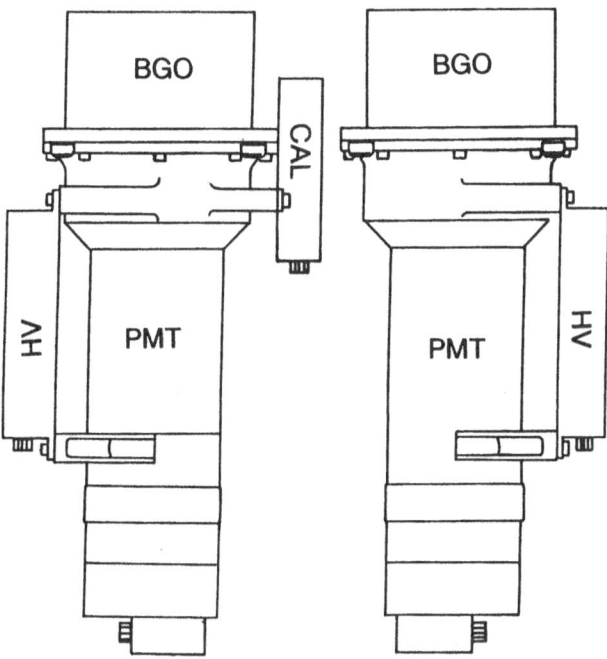

Fig. 14. Cross-sectional view of GRS. GRS consists of two identical BGO ($Bi_4Ge_3O_{12}$) scintillators (7.6 cm diameter and 5.0 cm thickness) optically coupled to photomultiplier tube (PMT). Each PMT is biased by high-voltage power supply unit (HV). CAL is an in-flight energy calibration device used for two BGO detectors.

pulse height data every 4 s (GRS-PHH1 and GRS-PHH2) and 2-channel pulse count data every 0.5 s (GRS-PC15 and 25 for 8–30 MeV and GRS-PC16 and 26 for 30–100 MeV). The in-flight energy calibration in GRS-L is achieved by the detection of 1.17 and 1.33 MeV lines from ^{60}Co nuclear de-excitation in coincidence with a β-ray (maximum energy 310 keV) detected with two Si detectors. The GRS in-flight energy calibration spectrum measured with the coincidence method is shown in Figure 20.

GRS also has detection sensitivity to solar neutrons. As an example, the GRS response to 45 MeV neutrons produced by $p(Li, Be)n$ reaction is shown in Figure 21. Although the GRS has sensitivity to neutrons, it is difficult to determine the incident neutron energy from the GRS pulse height spectrum.

2.4. RADIATION BELT MONITOR (RBM)

The RBM consists of two different detectors, NaI scintillation detector (5.1 cm in diameter and 1 cm in thickness) and Si detector (4 mm in diameter and 100 μm in thickness). These two detectors are pointed in the direction perpendicular to the Sun and are insensitive to solar flare X-rays and gamma-rays. The NaI detector produces

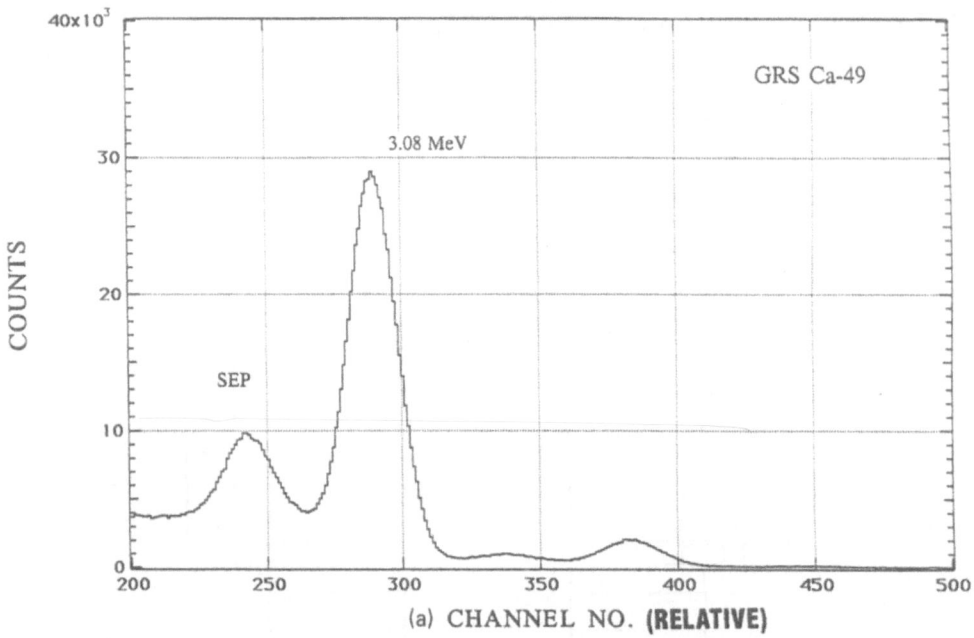

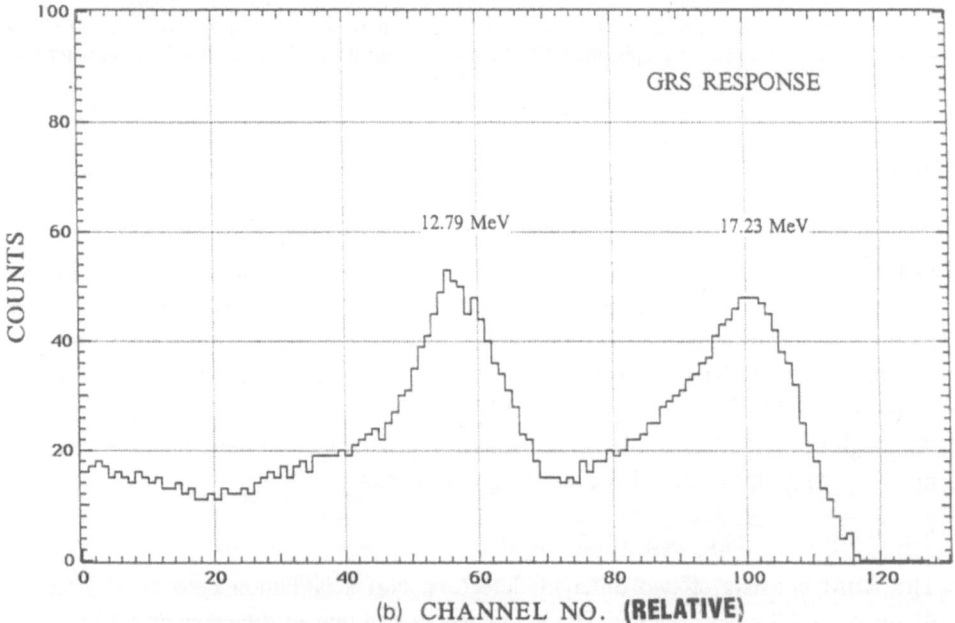

Fig. 15.   (a) GRS spectral response to 3.08 and 4.07 MeV lines from ^{49}Ca nuclear de-excitation. (b) GRS spectral response to 12.79 and 17.23 MeV lines from ^{7}Be nuclear de-excitation.

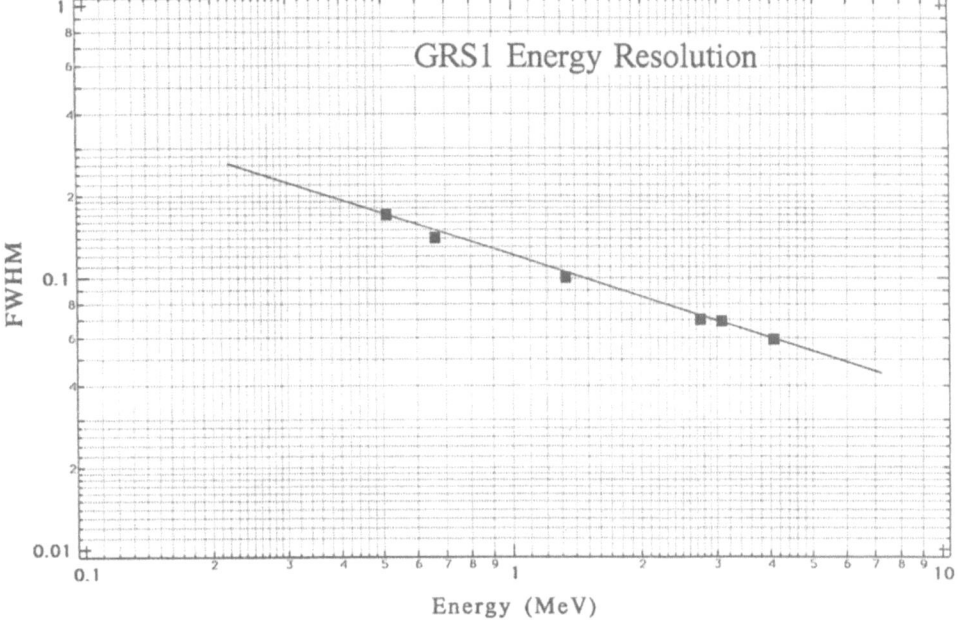

Fig. 16.   Energy resolution (FWHM) of GRS as a function of energy.

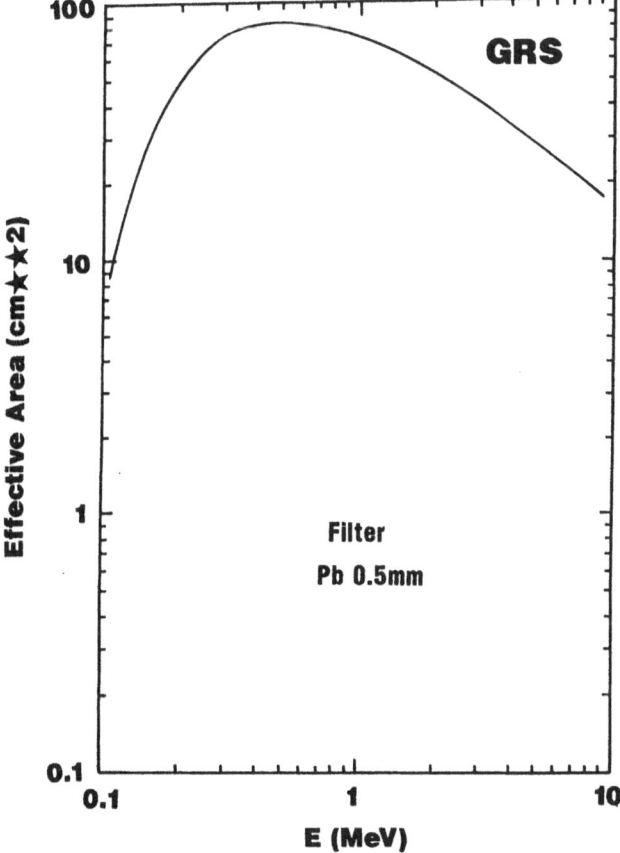

Fig. 17.   Calculated effective area of GRS as a function of energy.

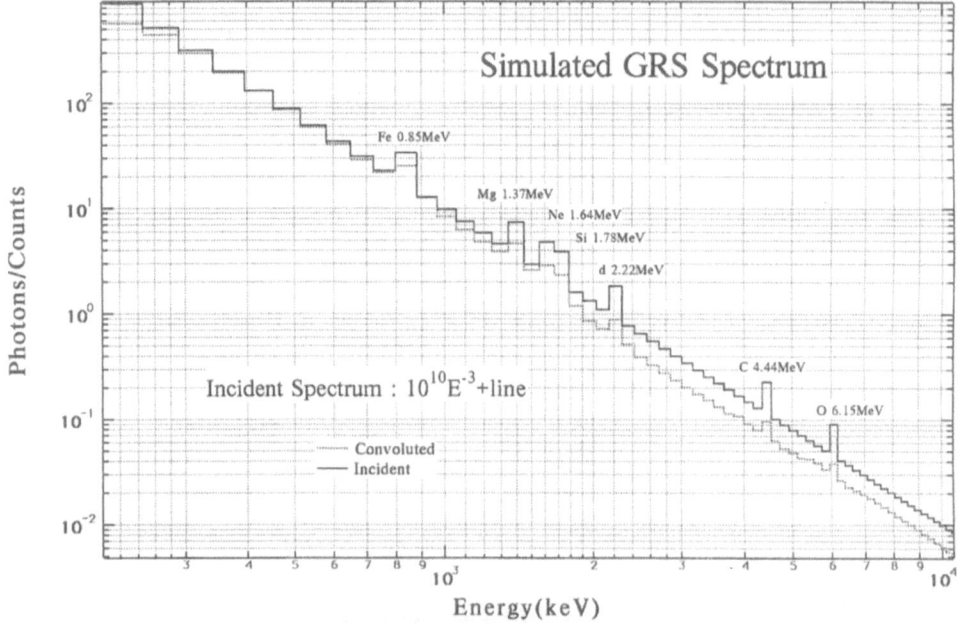

Fig. 18. Gamma-ray incident spectrum (E^{-3} + lines) and convoluted energy-loss spectrum.

32-channel quadratic-spaced pulse height data in the 5–300 keV range every 1 s and 2-channel pulse count data (5–60, 60–300 keV) every 0.25 s. The Si detector produces single channel pulse count data in energies above 20 keV every 0.25 s. When the pulse count of the NaI or the Si detector exceeds a certain threshold value, the alarm for the South Atlantic Anomaly (SAA) passage is sounded. The radiation belt alarm is used to avoid recording a false flare event.

2.5. GAMMA-RAY BURST DETECTION CAPABILITY

The HXS and RMB NaI detectors are capable of monitoring non-solar X-ray and gamma-ray burst phenomena during solar quiet time and night time. When the HXS

Fig. 19. GRS electronic block diagram. GRS consists of two identical BGO ($Bi_4Ge_3O_{12}$) scintillation detectors (GRS-1 and GRS-2). Each detector is connected to a similar electronic circuit and produces two outputs. One is an anode output pulse (0.2–10 MeV) and the other is a 6th dynode output pulse (8–100 MeV). The anode pulse passes through a preamplifier and shaping amplifier, and is fed to a main amplifier. After amplification, the output pulse is sent to an 8-bit ADC to produce 128-channel pulse height data (GRS-PHL). The amplified pulse is also sent to a discriminator with 5 levels to produce 4-channel pulse count data (GRS-PC11, 12, 13, and 14 for GRS-1 and GRS-PC21, 22, 23, and 24 for GRS-2). The dynode pulse is similarly processed to produce 16-channel pulse height data (GRS-PH) and 2-channel pulse count data (GRS-PC25 and 26 for GRS-2). Two silicon solid state detectors (SSD) output is fed to a preamplifier and shaping amplifier, and then sent to a discriminator. A coincidence event between the BGO and SSD detectors provides an in-flight energy calibration event. In addition, ADC triggering event (GRS-ADCLT and GRS-ADCHT), pulse count above upper discrimination level (GRS-UD), and SSD pulse count (GRS-CAL) are monitored as housekeeping data. HV level, main amplifier gain, and SSD discrimination level are changeable by block command (BC).

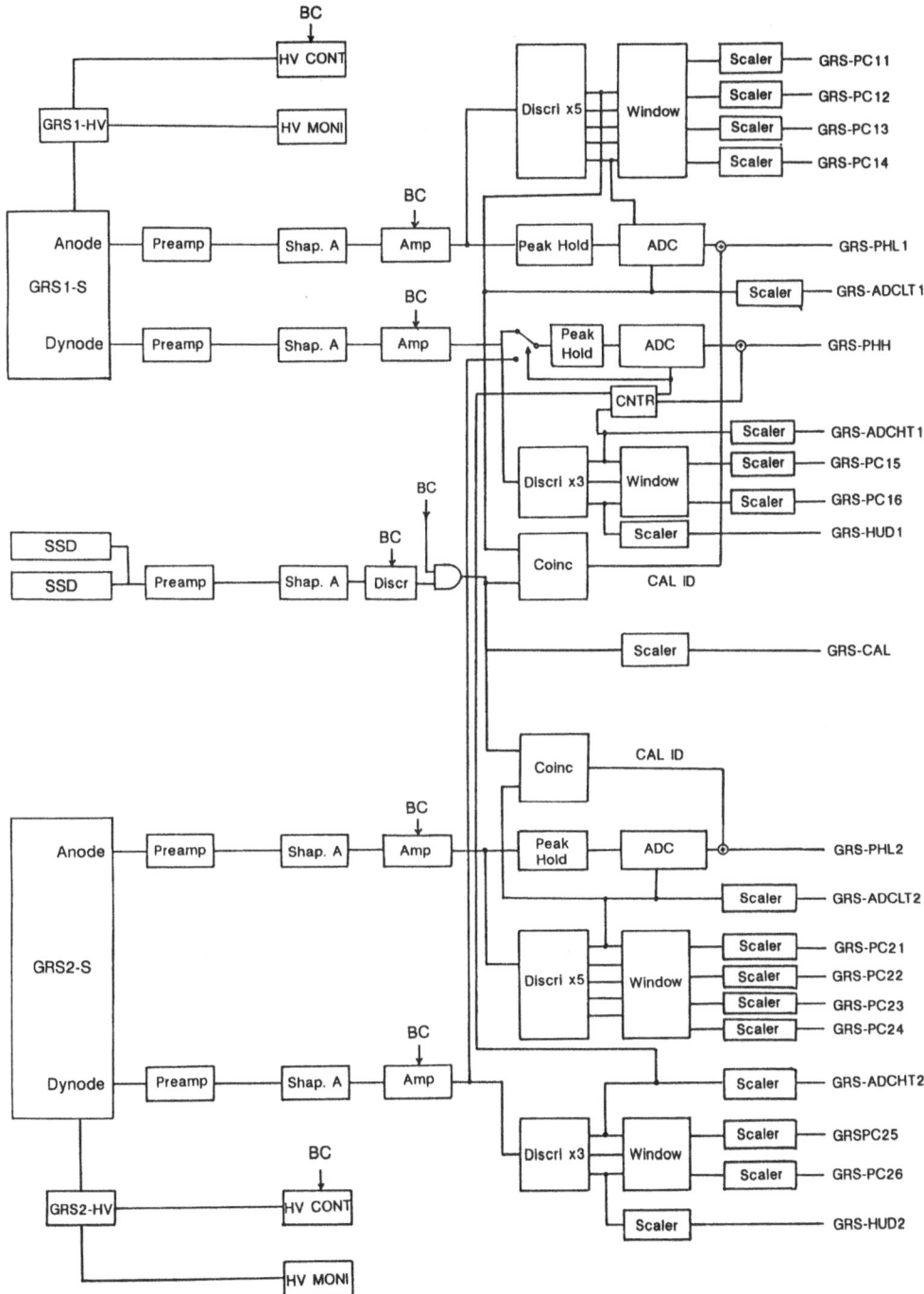

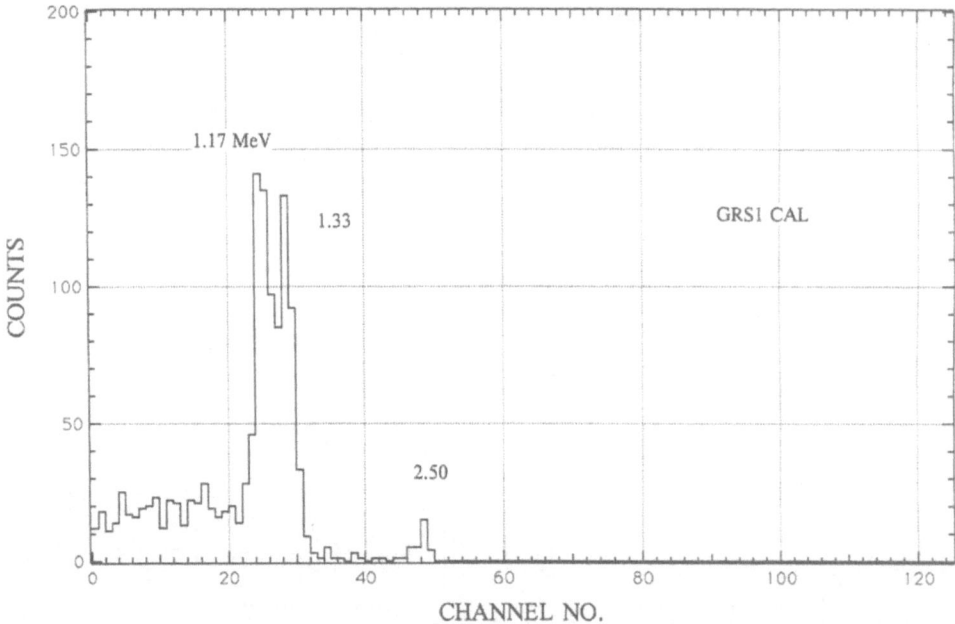

Fig. 20.   GRS in-flight energy calibration spectrum of 1.17 and 1.33 MeV lines from ^{60}Co nuclear de-excitation.

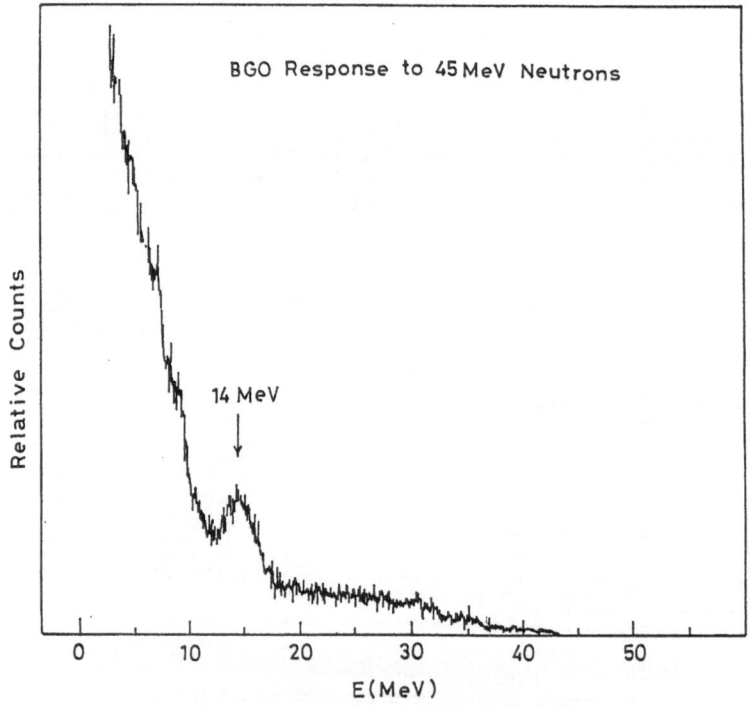

Fig. 21.   GRS response to 45 MeV neutrons.

pulse count in the 50–600 keV range or the RBM pulse count in the 60–300 keV range exceeds a certain threshold value, both 32-channel pulse height data every 1 s and pulse count data every 0.25 s are recorded for 256 s.

TABLE I

Principal WBS output data. SXS consists of two different gas proportional counters (SXS-1 and SXS-2) and each counter provides 128-channel pulse height (PH) data every 2 s, and 2-channel pulse count (PC) data every 0.25 s. HXS consists of a NaI scintillation detector and provides 32-channel PH data every 1 s, and 2-channel PC data every 0.125 s. GRS consists of two identical BGO scintillation detectors (GRS-1 and GRS-2) and each detector provides 128-channel (0.2–10 MeV) and 16-channel (8–100 MeV) PH data every 4 s, and 6-channel PC data every 0.25 or 0.5 s.

Spectrometer	Output	PH-data (time resolution)	PC-data (time resolution)
SXS	SXS-1	128-ch (2–30 keV) (2 s)	2-ch (0.25 s)
	SXS-2	128-ch (2–30 keV) (2 s)	2-ch (0.25 s)
HXS	HXS	32-ch (20–600 keV) (1 s)	2-ch (0.125 s)
GRS	GRS-1	128-ch (0.2–10 MeV) (4 s)	4-ch (0.25 s)
		16-ch (8–100 MeV) (4 s)	2-ch (0.5 s)
	GRS-2	128-ch (0.2–10 MeV) (4 s)	4-ch (0.25 s)
		16-ch (8–100 MeV) (4 s)	2-ch (0.5 s)

The principal WBS data are given in Table I. In addition to the principal data, housekeeping data are also recorded. These data include command status, instrument status, sixteen digital housekeeping data (the counts above upper energy discrimination level, ADC triggering counts, and inflight energy calibration event counts) and nine analog housekeeping data (high voltage and temperature).

3. Expected Scientific Return

WBS is expected to produce the following results. From the SXS observation the time development of electron temperatures and emission measures of a heated plasma can be studied and the cutoff energy of the nonthermal electron spectrum will be determined. The SXS observations can contribute to solving some fundamental problems of plasma heating during solar flares. From the HXS and GRS observations the electron and ion acceleration processes can be characterized in greater detail and the physical conditions which dominate the time scale and the efficiency of the particle acceleration mechanisms can be better determined. The WBS observations will provide an essential key to our understanding of high-energy processes such as plasma heating, particle acceleration, and interaction processes during solar flares.

Furthermore, collaborations with the other instruments of SOLAR-A (soft X-ray telescope, hard X-ray telescope, and Bragg crystal spectrometer), gamma-ray instruments on GRO and GRANAT, Max '91 HIREGS (high-resolution gamma-ray spectrometer) and HEIDI (high-energy imaging device), radio imagers at microwave and millimeter wavelengths, solar energetic particle (SEP) instruments on GEOTAIL and WIND, and ground-based solar neutron monitors will provide very fruitful and exciting results which may help us to solve a lot of questions related to the high-energy solar flare phenomena.

Acknowledgement

The authors wish to thank Toshiba Corporation for the detailed design, fabrication and testing of the electronics and for the integration of the WBS system.

Note Added in Proof

The HXS in-flight energy calibration method described here is not adopted. The HXS in-flight energy calibration is achieved by detection of 60 keV line of the attached Am-241 calibration source.

References

Chupp, E. L.: 1984, *Ann. Rev. Astron. Astrophys.* **22**, 359.
Chupp, E. L.: 1987, *Phys. Scripta* **T18**, 5.
Dennis, B. R.: 1988, *Solar Phys.* **118**, 49.
Forrest, D. J.: 1990, *21st Int. Cosmic Ray Conf.* **7**, 241.
Kondo, I., Yoshimori, M., Okudaira, K., Hirasima, Y., Igarashi, T., Akasaka, M., Nishimura, J., Yamagami, T., Ohki, K., and Watanabe, T.: 1990, *21st Int. Cosmic Ray Conf.* **7**, 268.
Watanabe, T., Tanaka, K., Akita, K., and Nitta, N.: 1983, *Solar Phys.* **83**, 107.
Yoshimori, M.: 1988, in N. Gehrels and G. H. Share (eds.), *Nuclear Spectroscopy of Astrophysical Sources*, AIP Conf. Proc. No. 170, p. 401.
Yoshimori, M.: 1989, *Space Sci. Rev.* **41**, 85.
Yoshimori, M. and Okudaira, K.: 1988, *Nucl. Instr. Meth.* **A272**, 880.
Yoshimori, M., Okudaira, K., Hirasima, Y., Yanagimachi, T., Kondo, I., Ohki, K., Watanabe, T., Nishimura, J., Yamagami, T., Murakami, T., Ito, M., and Yoshida, A.: 1988, *Nucl. Instr. Meth.* **A264**, 436.

THE BRAGG CRYSTAL SPECTROMETER FOR SOLAR-A*

J. L. CULHANE[1], E. HIEI[2], G. A. DOSCHEK[3], A. M. CRUISE[4],
Y. OGAWARA[5], Y. UCHIDA[6], R. D. BENTLEY[1], C. M. BROWN[3], J. LANG[4],
T. WATANABE[2], J. A. BOWLES[1], R. D. DESLATTES[7], U. FELDMAN[3],
A. FLUDRA[1], P. GUTTRIDGE[1], A. HENINS[7], J. LAPINGTON[1],
J. MAGRAW[4], J. T. MARISKA[3], J. PAYNE[4], K. J. H. PHILLIPS[4],
P. SHEATHER[1], K. SLATER[4], K. TANAKA[2]**, E. TOWNDROW[4],
M. W. TROW[1], and A. YAMAGUCHI[2]

Abstract. The Bragg Crystal Spectrometer (BCS) is one of the instruments which makes up the scientific payload of the SOLAR-A mission. The spectrometer employs four bent germanium crystals, views the whole Sun and observes the resonance line complexes of H-like Fe XXVI and He-like Fe XXV, Ca XIX, and S XV in four narrow wavelength ranges with a resolving power $(\lambda/\Delta\lambda)$ of between 3000 and 6000. The spectrometer has approaching ten times better sensitivity than that of previous instruments thus permitting a time resolution of better than 1 s to be achieved. The principal aim is the measurement of the properties of the 10 to 50 million K plasma created in solar flares with special emphasis on the heating and dynamics of the plasma during the impulsive phase. This paper summarizes the scientific objectives of the BCS and describes the design, characteristics, and performance of the spectrometers.

1. Introduction

The solar flare problem represents one of the most difficult challenges posed in astrophysics. In the past 15 years it has become clear that progress can best be made with the aid of observations throughout the widest electromagnetic spectrum. The Bragg Crystal Spectrometer (BCS) will be used at X-ray wavelengths to study plasma heating and dynamics particularly during the impulsive phase of solar flares. The emission lines selected for observation allow the measurement of plasma velocity, temperature, and emission measure. Study of these flare plasmas will be undertaken jointly by all the SOLAR-A instruments. The measurement of element abundances and abundance variations will be an important aim of BCS observations in particular.

High spectral and temporal resolution coupled with high sensitivity are necessary features of an instrument designed to achieve the above objectives. The BCS will have almost ten times greater sensitivity than was available from the instruments flown on SMM (Acton *et al.*, 1980), P78–1 (Doschek, 1983) and HINOTORI (Kondo, 1983) during the last solar maximum. It will employ fixed bent crystals with one-dimensional

[1] Mullard Space Science Laboratory, University College London, Holmbury, U.K.
[2] National Astronomical Observatory of Japan, Tokyo, Japan.
[3] US Naval Research Laboratory, Washington, D.C., U.S.A.
[4] Rutherford Appleton Laboratory, Abingdon, U.K.
[5] Institute of Space and Astronautical Science, Tokyo, Japan.
[6] University of Tokyo, Tokyo, Japan.
[7] National Institute for Standards and Technology, Washington, D.C., U.S.A.

* After the launch the name of SOLAR-A has been changed to YOHKOH.
** Tragically Professor K. Tanaka died on January 2, 1990.

Solar Physics **136**: 89–104, 1991.

position-sensitive proportional counters to register the spectra. The spectral resolution will be comparable to that of the similar instrument on SMM but the enhanced sensitivity will permit a time resolution of better than 1 s. A flexible on-board data processing system, including a large queue memory, will allow a wide range of operating modes to be implemented during the mission.

In the rest of the paper the scientific aims of the BCS are summarized, the design of the spectrometers is described with particular reference to the Bragg crystal performance, the X-ray detectors and the on-board processor. A brief account will be given of the instrument calibration. We must emphasize that the instrument performance parameters presented are preliminary in nature and require verification following *in-orbit* operation. They should not be used for data analysis. A users manual is being prepared and will be updated after the launch of SOLAR-A.

2. Scientific Objectives

Following the impulsive release of energy in a solar flare, a large quantity (emission measure $\simeq 10^{48}$–10^{50} cm^{-3}) of high temperature ($T \simeq 10$–50×10^6 K) plasma is created. The manner of the plasma creation and the dynamics and other properties of the high-temperature gas all provide essential clues to understanding the flare mechanism. In the above temperature range the high-ionization stages (e.g., H-like and He-like) of abundant elements are formed and the plasma properties may be studied by observing the X-ray emission spectra of these ions. The sensitivity and wavelength resolution of the spectrometers permit line intensities and profiles to be measured with sufficient time resolution to obtain detailed observations throughout the impulsive phase of flares. The data will be used in the following investigations:

Plasma dynamics. Observations with high-resolution X-ray spectrometers during the past decade have demonstrated that the emission lines are significantly broadened during the flare impulsive phase. The 'turbulent' width, substantially greater than the Doppler width implied by simultaneous electron temperature measurement, indicates mass motion velocities approaching 200 km s^{-1} (Doschek *et al.*, 1980; Antonucci *et al.*, 1982). There may also be a correlation of non-thermal line width with associated hard X-ray flux. In addition the broadening may be non-random on short time scales (Doyle and Bentley, 1987; Fludra *et al.*, 1989). A blue-shifted component is often observed for disc flares (Doschek *et al.*, 1980; Feldman *et al.*, 1980; Antonucci *et al.*, 1982) indicating upward moving plasma with velocity of 300 km s^{-1} or greater. This plasma has been attributed to 'chromospheric evaporation' – the ablation of plasma heated by energy transported form the flare site (Antonucci *et al.*, 1982; Antonucci, Gabriel, and Dennis, 1984) but the details of this process and of the energy transport mechanism, remain controversial. Use of the enhanced sensitivity and time resolution of the SOLAR-A BCS will lead to major advances in this important area of flare physics.

Plasma heating and diagnostics. There is significant evidence from previous observations that heating and turbulence occur even before the impulsive phase of the flare. The increased sensitivity of the BCS will permit early measurement of temperature with much increased statistical precision. It will also be possible to search for anomalous line ratios which will indicate transient ionization effects thus allowing the estimation of electron densities in the range below 10^{11} cm^{-3}. Greater sensitivity will permit improved estimation of plasma parameters such as temperature and emission measure. Comparison of these measurements with Soft X-ray Telescope (SXT) images which are expected to have better than 3 arc sec spatial resolution will allow the direct estimation of electron density and the determination of differential emission measure over a wide temperature range.

Superhot component. Observations of the H-like iron (Fe XXVI) emission line spectrum will be particularly important for the study of the plasma at temperatures of around 50×10^6 K that is detected in some flares (Lin *et al.*, 1981; Tanaka *et al.*, 1982). Temperatures will be obtained from the intensity ratio of the Fe XXV dielectronic lines to those of the Fe XXVI Lα lines. In addition it will be possible for the first time to obtain statistically significant profiles for the Lα lines and so investigate the dynamics of this important component. Data from HINOTORI (Tanaka, 1987) provided a tentative indication of variability in the Fe XXVI line profiles. The enhanced spectral resolution of the BCS will permit detailed observations of line profile variations with time.

Flare decay phase. Although the main release of energy in the flare occurs during the impulsive phase, there is evidence for continued energy injection during the decay phases of some flares. The BCS will permit a detailed study of this process. For large limb flares, it may be possible to observe the increase in height of the emitting plasma since spatial displacement translates into spectral displacement in the BCS. Such plasma, contained in large post-flare loops, was observed with the SMM BCS to have a temperature of $\simeq 5$–6×10^6 K many hours after the flare had occurred (Švestka *et al.*, 1982). Combined observations by the SXT and the BCS will allow a substantial advance in this area. Finally the measurement of line to continuum ratios during flare decay will allow flare-to-flare abundance variations to be studied for a range of elements (Sylwester, Lemen, and Mewe, 1984; Sylwester, 1987).

3. Design of the SOLAR-A Bragg Crystal Spectrometer System

The crystal spectrometers constructed for flight on SOLAR-A are similar to those successfully launched on the Solar Maximum Mission (Rapley *et al.*, 1977; Acton *et al.*, 1980). In the case of SOLAR-A however the presence of high-quality imaging X-ray telescopes, the smaller mass and volume available, and the results obtained by the SMM have led to a somewhat different approach to the instrument design.

Conventional Bragg spectrometers scan in wavelength by rotating a flat crystal so that a range of angular positions (θ) converts a range of wavelengths (λ) according to Bragg's law $n\lambda = 2 d \sin \theta$. For SOLAR-A each spectrometer crystal is curved with a fixed radius so that a parallel beam of solar X-rays is incident at a range of Bragg angles θ_1 to θ_2. Diffracted radiation, at corresponding wavelengths λ_1 to λ_2, is registered in a one-dimensional position-sensitive proportional counter (see Figure 1). Given the existence of the imaging X-ray telescopes, the small probability of simultaneous flare

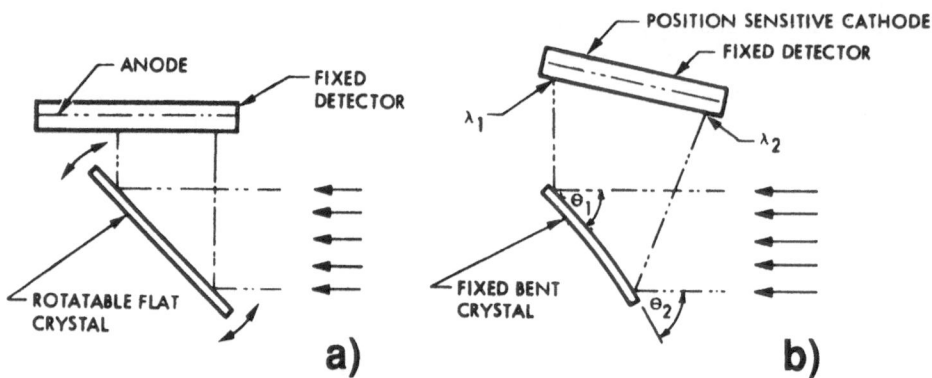

Fig. 1. Schematic diagram of (a) flat scanning crystal spectrometer and (b) fixed bent crystal spectrometer.

occurrence in different active regions and the relatively small mass and volume available for the BCS, a multi-grid collimator was not employed on SOLAR-A and so the BCS views the entire solar disc. This, coupled with increased crystal area, allows a factor approaching ten increase in sensitivity relative to the SMM instrument. The sensitivity of a bent crystal spectrometer can be obtained from the relation given by Rapley *et al.* (1977) as

$$S = [T_w T_f F_b \eta_d A_p R_c]/\Delta\theta, \tag{1}$$

where S is the sensitivity in cm^2, T_w and T_f are the transmissions of the detector window and the thermal filter, F_b is the fractional detector area lost due to the window support bars, η_d is the X-ray absorption efficiency of the detector gas, A_p is the projected area of the crystal in cm^2, R_c and $\Delta\theta$ are the integrated reflectivity of the crystal and the range of incident Bragg angles, both in radians.

From (1) S is inversely proportional to $\Delta\theta$ and therefore to the wavelength range $\lambda_2 - \lambda_1$ covered by each crystal. In order to maximize S it is therefore necessary to select the minimum wavelength range that is consistent with achieving the scientific aims. However, since Bragg angle range translates to wavelength range, the minimum spectral coverage must be increased to allow for uncertainties in spacecraft pointing, for spectrometer alignment errors, and for the range of flare locations on the surface of the Sun. With the crystal dispersion direction aligned approximately perpendicular to the solar equator, the allowance of an addtional ± 12 arc min in Bragg angle permits the chosen spectral ranges to be registered for all possible spacecraft pointing directions which lie inside a circle of 5 arc min angular radius about Sun centre and for most flares which occur mainly in a latitude range of ± 7 arc min about the equator.

The BCS employs four bent crystals to cover selected wavelength ranges of diagnostic importance. These ranges are shown in Figure 2 superimposed on spectra of Fe XXVI taken from HINOTORI data (Tanaka, 1987), Fe XXV and Ca XIX taken from SMM BCS data (Culhane *et al.*, 1981) and S XV taken from SMM Flat Crystal Spec-

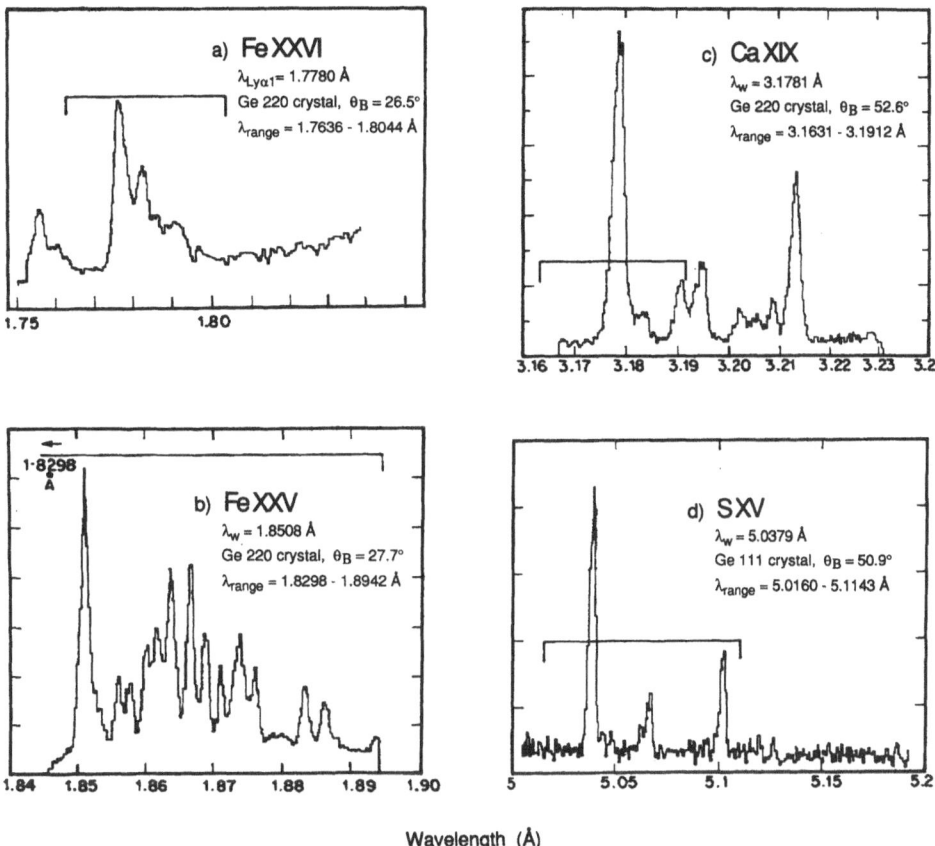

Wavelength (Å)

Fig. 2. Bragg crystal spectra of solar flares. (a) Fe XXVI obtained with HINOTORI, (b) Fe XXV and (c) Ca XIX obtained with the SMM bent crystal spectrometer, and (d) S XV obtained with the SMM flat crystal spectrometer. The solid lines indicate the spectral coverage for the four SOLAR-A spectrometers. See text for references.

trometer (FCS) data (Acton *et al.*, 1981). The Fe XXVI range will permit velocity and temperature measurements for the superhot component. Coverage of the Ca XIX range has been essentially restricted to the resonance transition to allow velocity measurements at maximum resolution and sensitivity. However, there is some temperature information available from the intensity of the $n = 3$ satellite lines that fall close to the long wavelength side of the resonance line. The more complete coverage of the Fe XXV and S XV ranges will permit both velocity and temperature measurements.

The four crystals are mounted in two structures as shown in Figure 3. X-rays from the Sun (Z-direction) enter the open apertures of the spectrometers through a pair of thin aluminized Kapton thermal filters (not shown in the figure) which are mounted over openings in the front panel of the spececraft. The incoming radiation strikes the crystals which are curved and fixed in place before launch so as to cover the selected wavelength ranges. The radiation is then diffracted into detectors located inside each structure. The

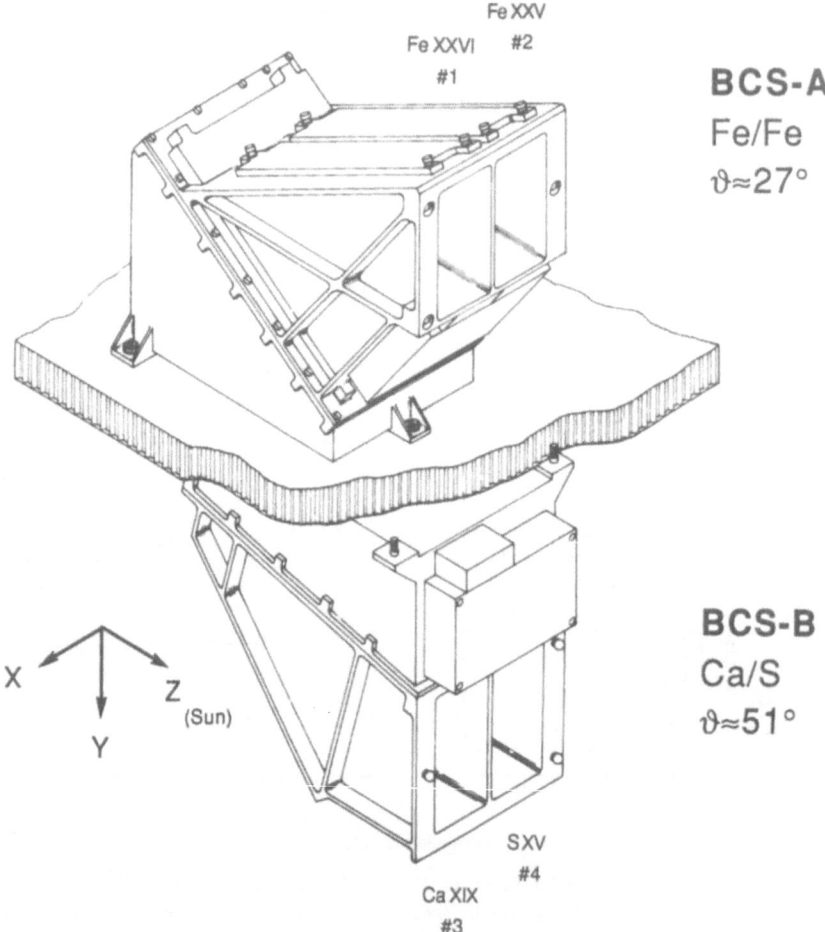

Fig. 3.   Isometric views of the two SOLAR-A spectrometers. The units are mounted on opposite sides of
the spacecraft centre panel.

only moving parts are two stepper motors which can rotate to admit 5.89 keV Mn–K
X-rays from Fe^{55} radioactive sources to the detectors to enable their gas gain and energy
resolution to be measured in flight.

Within the spectrometers an optical technique is used to set each crystal at the
appropriate Bragg angle with a precision of better than ± 15 arc sec with respect to the
plane defined by the feet of the spectrometers. Each spectrometer has three feet with
precisely located and sized screw holes to enable accurate mounting on the spacecraft
centre panel. Precision shoulder screws attach the feet to the centre panel. An optically
located drill template is used to position the mating holes in the centre panel in the
correct relationship to the Z-axis of the spacecraft. The openings in the centre panel
mounting holes are accurately reamed to accept the barrels of the shoulder screws.
Clearances of 25 μm are maintained so that the shoulder screws serve both as attach-

ments and as locating pins. The alignment procedure ensures that the optical axis of the BCS is co-aligned with the Z-axis of the spacecraft. As the centre panel is specified to be flat to within $\pm$ 3.4 arc min, provision is made for shimming the spectrometer during mounting on the spacecraft to optimize the Bragg angle ranges.

The two spectrometer units, labelled BCS-A and BCS-B, each contain two germanium crystals which have been paired so as to have approximately the same Bragg angles. The spectrometer structures are built with reference surfaces having the average of the two Bragg angles machined into the crystal support structure. Each crystal mount has three mounting pads that can be ground to ensure that the crystal offers the required range of Bragg angles to incoming solar radiation. The crystal mounts are made from titanium whose thermal expansion coefficient is well matched to that of germanium. The front surfaces of the mounts are ground to a concave cylindrical shape of the required radius. The 0.6 mm thick crystals are bent to the correct curvature and then attached to the mount with epoxy.

The crystal parameters and the wavelength ranges covered are listed in Table I. The crystals are somewhat larger than those flown on earlier missions. While Ge (220) crystals were used in both the P78–1 and SMM spectrometers, the Ge (111) crystal has not been flow previously. Before mounting in the spectrometer each crystal was characterized by measurements of its curvature and integrated reflectivity using X-rays.

The BCS instrument response is summarized in Table II. The wavelength resolution is set by the geometry of the spectrometers, the crystal rocking curves and the X-ray position resolution of the detectors. Each detector has a single position encoding arrangement – a modified wedge and strip pattern. X-rays from pairs of crystals are registered in each detector and are identified by signals from separate anode wires. The operation and performance of the detectors is dealt with in the next section.

Spectra are accumulated in the on-board digital processing system for a fixed time interval after which the total count in each spectral bin is compressed to an 8-bit number. At the flare-mode telemetry rate of 2 kbits s^{-1} for the BCS, four complete spectra can be transmitted every 4 s. However, the digital system incorporates a large queue memory which can be filled with spectra at a faster rate for later transmission. For example, the spectrometer can store a series of four spectra with 1 s time resolution for up to 7 min thus acquiring a total of 420 sets each of four spectra. This mode of operation can be initiated by an on-board flare occurrence flag thus enabling spectra to be acquired at higher time resolution during the impulsive phase. The digital data system will be described more fully in Section 5.

4. Detectors and Analogue Electronics

An exploded view of the one-dimensional position-sensitive proportional counter is given in Figure 4. The detector is filled with a mixture of Ar and Xe both at 47.5% with CO_2 (5%) as a quench gas to an overall pressure of 1.2 atm. The body is in two stainless steel halves. A beryllium foil window of 125 μm thickness is brazed to the upper section while the anodes, cathodes, and readout pattern are installed in the lower section. The

TABLE I

Ions, crystal parameters, and wavelength ranges

Channel No.	Ion	Resonance line λ (Å) (Bragg angle)	$2d$ (Å)	Crystal (rocking curve)	Crystal bend radius (m)	Wavelength range (Å) ($\Delta\theta$, mrad)	Crystal size (cm)	R_c (µrad)
BCS-A								
1	Fe XXVI	1.7780 (26.39°)	4.000	Ge 220 (14 arc sec)	13.64	1.7636–1.8044 (11.40)	3.98 × 18.1	67
2	Fe XXV	1.8509 (27.56°)	4.000	Ge 220 (15 arc sec)	10.20	1.8298–1.8942 (18.19)	3.98 × 18.1	68
BCS-B								
3	Ca XIX	3.1769 (52.58°)	4.000	G2 220 (50 arc sec)	9.60	3.1631–3.1912 (11.57)	3.98 × 11.4	116
4	S XV	5.0385 (50.46°)	6.532	Ge 111 (68 arc sec)	4.56	5.0160–5.1143 (23.83)	3.98 × 11.4	121

TABLE II

BCS instrument response

Channel No.	Ion	Wavelength range (Å)	Wavelength resolution (mÅ)	Wavelength resolution ($\lambda/\Delta\lambda$)	Thermal Doppler width FWHM (mÅ)	T_{ion} (10^6 K)	Sensitivity S (cm^2)	SOLAR-A/SMM
BCS-A								
1	Fe XXVI	1.7636–1.8044	0.38	4700	1.20	50	0.15	9 [a]
2	Fe XXV	1.8298–1.8942	0.53	3500	0.90	25	0.10	9 [a]
BCS-B								
3	Ca XIX	3.1631–3.1912	0.53	6000	1.60	20	0.20	6 [a]
4	S XV	5.0160–5.1143	1.86	2700	2.50	15	0.04	63 [b]

[a] SMM bent crystal spectrometer.
[b] SMM flat crystal spectrometer scanning the S XV resonance line at 10 arc sec.

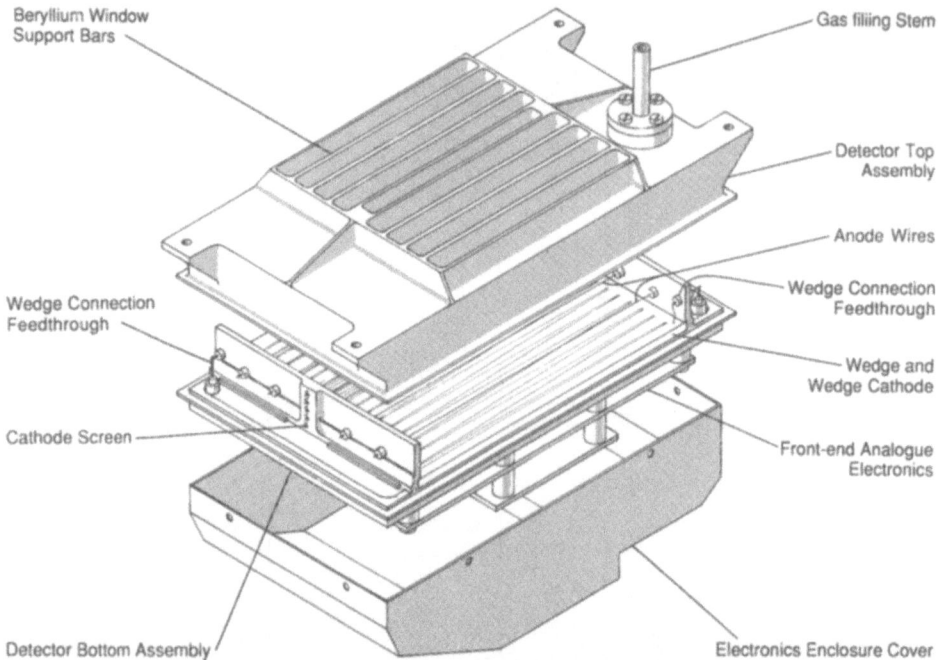

Beryllium Window
Support Bars

Gas filling Stem

Detector Top
Assembly

Anode Wires

Wedge Connection
Feedthrough

Wedge Connection
Feedthrough

Wedge and
Wedge Cathode

Cathode Screen

Front-end Analogue
Electronics

Detector Bottom Assembly

Electronics Enclosure Cover

Fig. 4.   An exploded view of the double proportional counter. A single one-dimensional position readout pattern is common to both halves. Each half detector registers X-rays from a single crystal and has two connected anode wires. The two halves are electrically separated by a screen of cathode wires. The top and bottom assemblies are sealed together by electron beam welding.

two sections are then sealed together by electron beam welding. The detector has two pairs of 15 μm dia. anode wires separated by a cathode screen of 9 grounded 25 μm dia. wires. Each pair of anodes is connected to a separate pre-amplifier. Thus X-rays from each of the two crystals can be registered separately. Anodes are maintained at a potential of ∼ 1.5 kV by a high voltage unit (HVU). In order to monitor the gain and energy resolution of each detector, a 32-channel diagnostic pulse-height analyzer (PHA) is available (Figure 5) whose input can be switched to any one of the four detector channels. Used in conjunction with the on-board Fe^{55} calibration sources, the PHA allows detector performance to be checked during spacecraft night.

The double-wedge readout pattern is shown schematically in Figure 5. It is manufactured photolithographically on a gold-coated fused silica plate. The substrate material is chosen for low dielectric constant so as to minimize the capacitance between the two sets of gold wedge electrodes. This ensures low noise in the wedge 1 (W1) and wedge 2 (W2) pre-amplifiers. Following the absorption of an X-ray photon, a charge avalanche occurs on one of the anode wires and the induced charge distribution on the two interlocking wedge electrodes provides the event position determination. If Q_{W1} and Q_{W2} are the induced charges on each electrode then the one-dimensional position

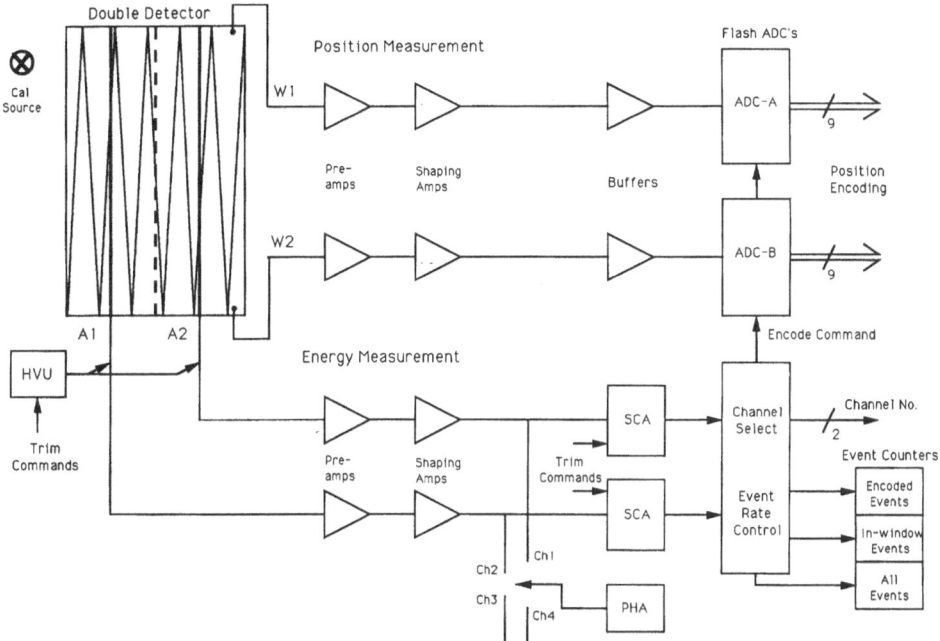

Fig. 5.   A schematic diagram of the X-ray event processing electronics. Signals from the two wedge patterns (W1 and W2) are amplified and passed to 9 bit analogue-to-digital converters. Anode signals (A1 and A2) from the double detector are amplified and passed to single channel analysers (SCA). A switch-selectable pulse height analyser (PHA) can examine the X-ray events from each anode of the two double detectors. Each detector has its own high voltage unit (HVU).

coordinate for the X-ray event is given by

$$x = Q_{w1}/(Q_{w1} + Q_{w2}).$$ (2)

The induced charge pulses Q_{w1} and Q_{w2} are digitized in two ADC's (Figure 5) and the value of x is determined by means of a look-up table (Figure 6). Events are ascribed to the appropriate crystal depending from which anode preamplifier (A1 or A2) the pulse originated. Since hard solar X-rays can cause the germanium crystals to emit 9.9 keV fluorescence radiation, single-channel pulse analyzers are employed to define windows which can accept photons from the crystals (E_{max} = 6.9 keV for Fe xxvi) while rejecting the germanium fluorescence photons. A position resolution of 350 μm FWHM together with a photon energy resolution of 17% has been achieved with the SOLAR-A detectors. The former ensures good spectral resolution, while in the Fe xxvi and Fe xxv channels the uniformity of gas gain over the detector window to better than 5% allows a reduction of a factor 500 or more in the rate of processed fluorescence photons for the loss of less than 2% of the diffracted X-rays.

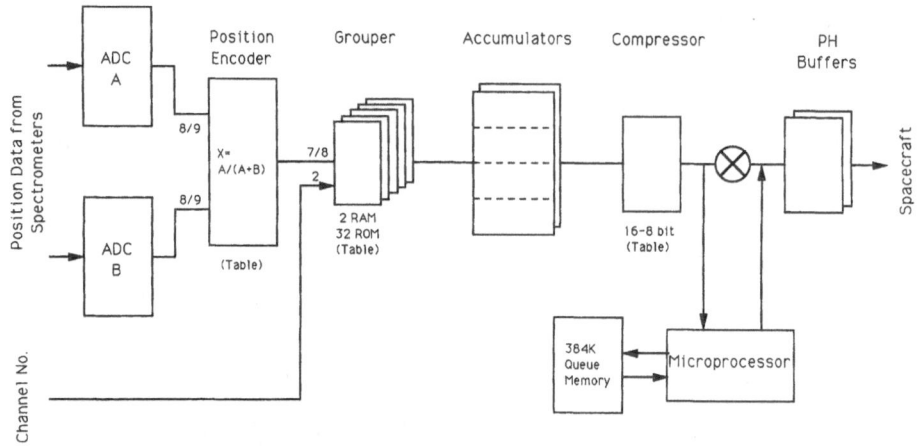

Fig. 6. The on-board data processing system. Event positions are determined in a look-up table. Spectra are assembled in the accumulators in accordance with one of the plans contained in the data grouper. The number of events in each spectral channel is compressed to an 8-bit value. Spectra can be stored in a queue memory or passed directly to buffers for the main digital data (PH) interface with the spacecraft.

5. The On-Board Processor and Data Handling

The need to acquire spectral data with good time resolution and to make optimum use of a relatively small telemetry allocation has led to the use of a sophisticated on-board data acquisition and control system which is illustrated schematically in Figure 6. Following each detected X-ray event, a pair of outputs, Q_{w1} and Q_{w2} of either 8 or 9 bits, is presented to the position encoder which is implemented as a look-up table. Using the 8-bit wavelength bin address produced by the position encoder (x in Equation (2)) together with information as to the channel of origin, an event is integrated in the accumulator. This consists of two buffers each of (4×256) 16-bit deep wavelength bins. This double buffer system allows data acquisition in one buffer while the other is being read out. The accumulators are sized to accommodate up to 256 bins in each channel although only 128 bins will normally be used for the Ca XIX and S XV channels. In order to optimize the number of wavelength bins used, before the event is deposited in the accumulator it is re-grouped into a 'smaller' number of bins through the use of a data grouper which is also implemented by a look-up table. Several possible grouping plans are always selectable. After the data has been accumulated for an integration period controlled by the accumulator timer, the accumulator buffers are swapped and the data is transferred from the accumulator to a hardware data compressor which is again implemented by a look-up table. Here the accumulated value of each 16-bit wavelength bin is reduced to an eight bit value before being stored in the data queue memory by the microprocessor.

Event data is output through the BCS-PH or main digital data interface to the spacecraft data processor in 256 byte blocks. The rate at which data is transferred by the interface is purely a function of the overall telemetry rate. The 384 kbyte data queue memory allows the BCS to produce bursts of spectral data for storage at rates that are in excess of the 2 kbits s^{-1} accepted by the spacecraft telemetry system in flare mode. The amount of spectral data produced is a function of the selected data grouping plan and the accumulator integration time. The microprocessor can change the values of these parameters as a flare develops to allow a trade-off between temporal and spectral resolutions through the flare. The criteria for switching between accumulation modes and the definition of the modes themselves are contained in lists that can be loaded into the microprocessor. Since the use of the queue memory makes the PH data asynchronous to the rest of telemetry, each accumulator block in the queue is headed by a block which contains the start time and other important information.

Even rate counters (Figure 5) driven from different parts of the event control circuitry are available in the housekeeping data stream. The counters report 'total', 'in-window', and 'encoded' photons for each channel and can be used to apply deadtime corrections when the countrates are high. They are also used by the microprocessor to watch for flares and to monitor the background rate. The diagnostic PHA data used to analyze the detector performance and a field containing information generated by the microprocessor are also contained within the housekeeping data stream.

BCS electronic subsystems such as high voltage units (HVU) and single-channel analyzers (SCA) are controlled through the microprocessor data bus. Parameters may be loaded through this bus, either directly from the command interface or from the microprocessor. The value of these parameters and other information on the status of relays and switched circuitry, together with the last two received command bytes, is given in the status data stream. Structure temperatures and HV monitors are measured through the analogue data interface. In the event of a hardware problem with the microprocessor, a backup path has been provided by which spectral data flows directly from the data compressor to the BCS-PH interface. In this mode, since the data queue is not available, the number of wavelength bins produced per second must be matched to the telemetry rate by selecting an appropriate combination of the data grouping plan and accumulator integration time.

6. Instrument Calibration

Although elements of the spectrometers were calibrated and tested individually, each spectrometer was also tested end-to-end when finally assembled for flight. These tests were designed to provide a wavelength calibration and an intensity calibration at one wavelength for each of the four BCS channels.

The apparatus is shown schematically in Figure 7. The X-ray source was used to illuminate a monolithic channel-cut crystal (Quartz 1011, $2d = 6.952$ Å) which was placed about 0.8 m from the source. The channel-cut crystal was used as a two-crystal monochomator with the crystals in the $(1, -1)$ setting. Four narrow slits, two between the X-ray source and the channel cut crystal and two between the channel-cut crystal and the test chamber served to limit the beam divergence and to select the output wavelength range of the monochromator. For source radiation incident on the crystal

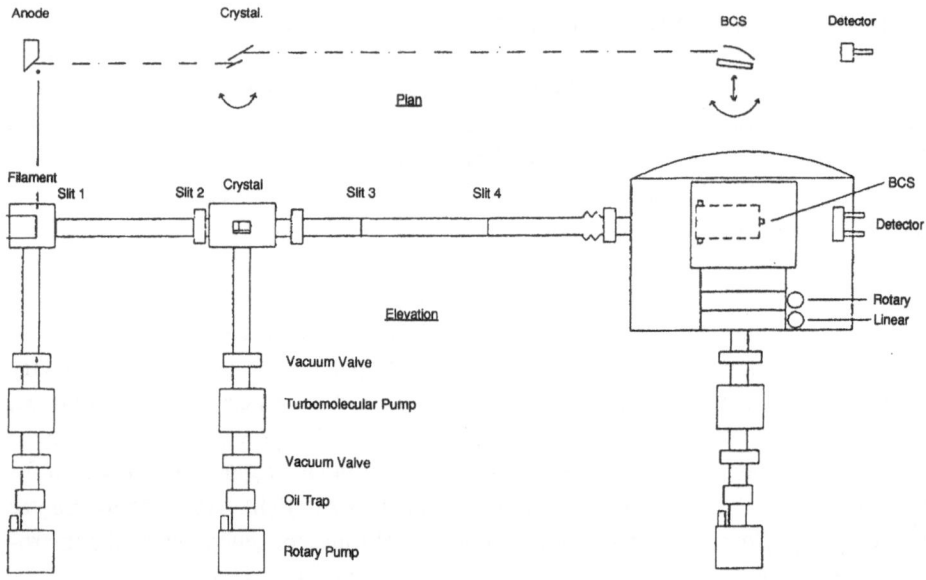

Fig. 7.   The end-to-end X-ray calibration system.

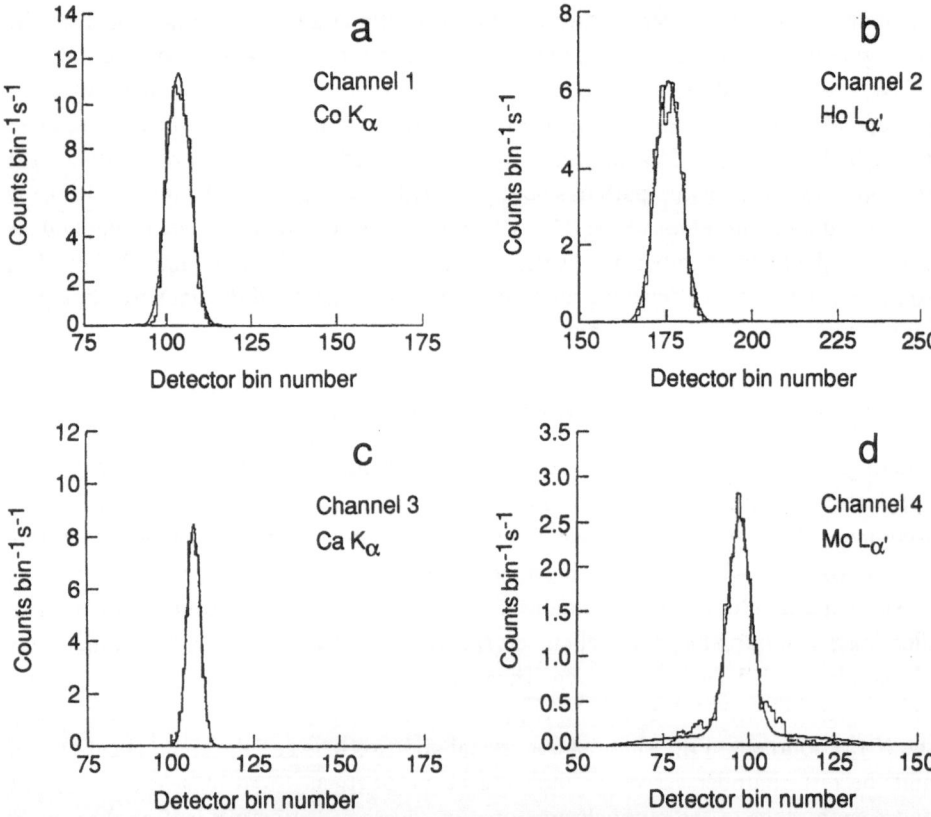

Fig. 8.   Results of the end-to-end X-ray calibration of the BCS. Spectra of (a) Co Kα¹ in the Fe XXVI channel (Ch. 1), (b) Ho Lα¹ in the Fe XXV channel (Ch. 2), (c) Ca Kα² in the Ca XIX channel (Ch. 3), and (d) Mo Lα¹ in the S XV channel (Ch. 4).

at the correct Bragg angle for a particular Kα line, the output had the line wavelength as seen by the BCS with a profile characterized by the source linewidth, the double crystal rocking curve of the monolithic crystal and the geometry of the slits. When measured using BCS-A by rotating the instrument, the line profile had a full-width-at-half-maximum (FWHM) of around 70 arc sec for the Fe XXVI channel. The channel-cut crystal was mounted in a holder whose position was adjusted to allow changes of Bragg angle for different source wavelengths. Spectrometer units were placed about 2 m from the double crystal. The precision ground feet of each spectrometer were fixed to a vertical surface which could be moved horizontally. With the BCS aligned correctly to an X-ray beam of appropriate and known wavelength, spectral calibrations were obtained with a precision ranging from 5 mÅ for the S XV channel to better than 1 mÅ for the Fe XXVI channel. The intensity calibrations were also established with a precision of $\pm 15\%$ using a proportional counter of known quantum efficiency and geometry to interrupt the beam incident on the BCS.

For the calibration of BCS-A, lines of Co Kα^1 at 6.930 keV (1.786 Å) and Ho Lα^1 at 6.720 keV (1.842 Å) were used for the Fe XXVI and Fe XXV channels, respectively. In the case of BCS-B it was necessary to rotate the spectrometer slightly in the plane of dispersion so that lines of Ca Kα^1 at 3.691 keV (3.354 Å) and Mo Lα^1 at 2.293 keV (5.399 Å) could be used for the Ca XIX and S XV channels. The response of all four spectrometer channels to the calibration lines is indicated in Figure 8.

Acknowledgements

We gratefully acknowledge the role of the Japanese Institute for Space and Astronautical Science (ISAS) who are responsible for the SOLAR-A mission and who made it possible for the BCS team to participate in this unique study of the solar flare phenomenon. We have had through all phases of the programme the help of many colleagues at the Mullard Space Science Laboratory, the National Astronomical Observatory of Japan, the US Naval Research Laboratory (NRL), the Rutherford Appleton Laboratory, the US National Institute for Standards and Technology and of course ISAS. To all of them we give our sincere thanks. The UK groups acknowledge the support of the Science and Engineering Research Council and the British National Space Centre. The Japanese activities were made possible by the support of the grant-in-aid for International Scientific Research Program No. 63044163 and No. 01044044 of the Japanese Ministry of Education, Science and Culture. Work in the U.S.A. was supported by the US Naval Research Laboratory.

References

Acton, L. W., Culhane, J. L., Gabriel, A. H., Bentley, R. D., Bowles, J. A., Firth, J. G., Finch, M. L., Gilbreth, C. W., Guttridge, P., Hayes, R. W., Joki, E. G., Jones, B. B., Kent, B. J., Leibacher, J. W., Nobles, R. A., Patrick, T. J., Phillips, K. J. H., Rapley, C. G., Sheather, P. H., Sherman, J. C., Stark, J. P., Springer, L. A., Turner, R. F., and Wolfson, C. J.: 1980, *Solar Phys.* **65**, 53.

Acton, L. W., Culhane, J. L., Gabriel, A. H., Wolfson, C. J., Rapley, C. J., Phillips, K. J. H., Antonucci, E., Bentley, R. D., Hayes, R. W., Joki, E. G., Jordan, C., Kayat, M., Kent, B. J., Leibacher, J. W., Nobles, R. A., Parmar, A. N., Strong, K. T., and Veck, N. J.: 1981, *Astrophys. J.* **244**, L137.

Antonucci, E., Gabriel, A. H., and Dennis, B. R.: 1984, *Astrophys. J.* **287**, 917.

Antonucci, E., Gabriel, A. H., Acton, L. W., Culhane, J. L., Doyle, C. J., Leibacher, J. W., Machado, M. E., Orwig, L. E., and Rapley, C. G.: 1982, *Solar Phys.* **78**, 107.

Culhane, J. L., Gabriel, A. H., Acton, L. W., Rapley, C. G., Phillips, K. J. H., Wolfson, C. J., Antonucci, E., Bentley, R. D., Catura, R. C., Jordan, C., Kayat, M., Leibacher, J. W., McWhirter, P., Parmar, A. N., Sherman, J., Springer, L. A., Strong, K. T., and Veck, N. J.: 1981, *Astrophys. J.* **244**, L141.

Doschek, G. A.: 1983, *Solar Phys.* **86**, 9.

Doschek, G. A., Feldman, U., Kreplin, R. W., and Cohen, L.: 1980, *Astrophys. J.* **239**, 725.

Doyle, J. G. and Bentley, R. D.: 1987, *Astron. Astrophys.* **155**, 278.

Feldman, U., Doschek, G. A., Kreplin, R. W., and Mariska, J. T.: 1980, *Astrophys. J.* **241**, 1175.

Fludra, A., Lemen, J. R., Jakimiec, J., Bentley, R. D., and Sylwester, J.: 1989, *Astrophys. J.* **344**, 991.

Kondo, I.: 1982, *Hinotori Symposium on Solar Flares*, ISAS, Tokyo, p. 3.

Lin, R. P., Schwartz, R. A., Pelling, R. M., and Hurley, K. C.: 1981, *Astrophys. J.* **251**, L109.

Rapley, C. G., Culhane, J. L., Acton, L. W., Catura, R. C., Joki, E. G., and Bakke, J. C.: 1977, *Rev. Sci. Inst.* **48**, 1123.

Švestka, Z., Stewart, R. T., Hoyng, P., Van Tend, W., Acton, L. W., Gabriel, A. H., Rapley, C. G., Boelee, A., Bruner, E. C., de Jager, C., Lafleur, H., Nelson, G., Simnett, G. M., Van Beek, H. F., and Wagner, W. J.: 1982, *Solar Phys.* **75**, 305.

Sylwester, J.: 1987, *Artificial Satellites/Space Phys.* **22**, 17.

Sylwester, J., Lemen, J. R., and Mewe, R.: 1984, *Nature* **310**, 665.

Tanaka, K.: 1987, *Publ. Astron. Soc. Japan* **39**, 1.

Tanaka, K., Watanabe, T., Nishi, K., and Akita, K.: 1982, *Astrophys. J.* **254**, L59.

SOLAR-A REFORMATTED DATA FILES AND OBSERVING LOG*

M. D. MORRISON, J. R. LEMEN, L. W. ACTON

Lockheed Palo Alto Research Laboratory, Palo Alto, CA 94304, U.S.A.

R. D. BENTLEY

Mullard Space Science Laboratory, University College London, Holmbury St. Mary, Dorking, Surrey, England

T. KOSUGI, S. TSUNETA

Institute of Astronomy, University of Tokyo, Mitaka, Tokyo 181, Japan

Y. OGAWARA

Institute of Space and Astronautical Sciences, Sagamihara, Japan

and

T. WATANABE

National Astronomical Observatory of Japan, Mitaka, Tokyo 181, Japan

Abstract. All of the SOLAR-A telemetry data will be reformatted before distribution to the analysis computers and the various users. This paper gives an overview of the files which will be created and the format and organization which the files will use. The organization has been chosen to be efficient in space, to ease access to the data, and to allow for the data to be transportable to different machines. An observing log file will be created automatically using the reformatted data files as the input. It will be possible to perform searches with the observing log to list cases where instruments are in certain modes and/or seeing certain signal levels.

1. Introduction

It is universally recognized that solar research, and especially flare studies, profit greatly from coordinated analysis of different types of observational data. Yet, in the past, little preplanning of data formatting and archiving has been done to facilitate joint analysis. Each experimenter tended to act independently, solving their own analysis and archiving problems in their own way. Data have been formatted and stored in a plethora of formats and media. Those data in digital form, and thus presumably universally available, reside in incompatible machines addressable only with specialized, esoteric software. Often the comparison of results has been seriously impeded. Joint analyses, other than the most qualitative comparisons, have been accomplished with great difficulty and in a sadly limited number of cases.

The SOLAR-A investigators have determined to do better. It has been decided to produce a common reformatted data base for all of the SOLAR-A experiments for

---

* After the launch the name of SOLAR-A has been changed to YOHKOH.

Solar Physics **136**: 105–110, 1991.
© 1991 *Kluwer Academic Publishers.*

distribution to all of the investigators. This will be produced with a synoptic file structure to ease access for scientific analysis and in a form as transportable as possible between computers with different operating systems. In the end, this work may be as important to the scientific productivity of the mission as the capabilities of the SOLAR-A instruments.

2. SOLAR-A Data Flow

The SOLAR-A data will be received by the Kagoshima Space Center (KSC) station in Japan, and by several NASA Deep Space Network (DSN) stations around the world (Canberra, Goldstone, and Madrid). The spacecraft uses a bubble data recorder (BDR) to store the data when a real-time station downlink is not available. The 10 Mbyte capacity of the BDR can store 40 min of high rate (32 kbps) data. There will be 5 downlinks to KSC every day plus four to ten downlinks to the DSN stations. The KSC data is available in real time; the DSN data could take up to two or three days before being available. All of the SOLAR-A raw telemetry data will be time-ordered and stored on-line in the large SIRIUS data base at the Institute of Space and Astronautical Sciences (ISAS). The maximum data telemetered down is 4.7 Gigabyte per month. The SOLAR-A reformatter will work from this data base.

The SOLAR-A reformatter will run on a Unix workstation at ISAS. The reformatter will be written in Interactive Data Language™ (IDL) Version 2.0 which is available from Research Systems, Inc. We intend to create data files that are portable and which can be read on any machine. The information necessary to tell the user how to read the file are present in the beginning of each data file. That information along with a copy of the SOLAR-A File Control Document should allow any user to access the data. The reformatter will create seven files as follows:

BCS	Bragg Crystal Spectrometer;
HXT	Hard X-ray Telescope;
SXT–PFI	Soft X-ray Telescope/Partial Frame Images;
SXT–FFI	Soft X-ray Telescope/Full Frame Images;
WBS	Wide-Band Spectrometer;
ATT	Spacecraft Attitude Data;
CBA	Spacecraft Common Basic Part.

The first five files contain the data from the scientific instruments. A complete copy of the raw spacecraft attitude data will be available in the ATT file. This will be the first file created and the reformatter will use this file to generate processed pointing information which will be included with each scientific data set entry. The CBA section contains a complete duplicate copy of the spacecraft common 'Basic Part' data. This is for possible unanticipated use in the future as we intend that all necessary information in this section of telemetry has been included in the respective scientific data files. The reformatted data files will initially be distributed on 8 mm tapes, about 1 such tape per week. We hope eventually to archive the data on CD-ROM.

3. Organization and Format of Reformatted Data Files

The data files are a simple stream of bytes, organized to allow direct access into the data file. The data is padded where necessary in order to package the data such that Vax VMS™ can use fixed records (probably 16 bytes long). Unix allows the user to directly access any byte and does not use record structures. This organization should allow us to maintain full compatibility between a large variety of machines.

The data files make full use of the structure data types which are available in FORTRAN (Vax and Sun™ FORTRAN extensions) and IDL. Each structure has a version number associated with it so structures can be changed and expanded without requiring difficult changes to the access software. Full flexibility to expand structures and add structures allows the user to maintain a full history of what processing has been done to the data. In addition to the raw reformatted data files, all processed data files will follow the same organization in order to allow existing software routines to display, analyze and process those new files.

The organization of the files is illustrated in Table I. Its sections will now be discussed in turn.

File pointer and information section. The first section in the file contains information needed to read the file. It contains pointers to the start of all of the major sections (file header, quasi-static index, data, optional data, and road map). It also contains information on what machine convention was used for integer and real variables, the file organization used, and the VMS record length. There are also two fields which hold sample integer and real values to confirm that any routine used for conversion between machine representations is working correctly.

File header section. The file header section contains information on what data is present in the file. Information about the range of dates and times covered, the program that created the file, the time and date that the data file was created, the total number of data sets, the machine that created the file, the spacecraft and instrument, the number of quasi-static index entries, and other similar information is contained in this section.

Quasi-static index section. The quasi-static index section is used to provide calibration and scientific data conversion information. The data present in this section seldom changes and does not need to be duplicated in each data index section. Information such as detector gain, coefficients for converting temperature and voltage, instrument pointing offsets, information on versions used for certain algorithms, etc., are included here.

Index and data section. The data are broken up into data sets, and each data set has an index section and a data section. The data are broken up by modes for SXT and BCS (WBS and HXT provide a continuous stream of data and do not have modes).

BCS	One data set per spectra.
HXT	One data set per telemetry major frame.
SXT	One data set per image exposure.
WBS	One data set per 2 telemetry major frames.

TABLE I
SOLAR-A reformatted files

	File Info / Pointer Section
	File Header
	Quasi-Static Index Section
Data Index	
Data	
Data Index	
Data	
	Index and Data Section
Data Index	
Data	
	Optional Data Section
	Road Map Section

The data index section will contain the information necessary to analyze the data. Information on the data word type (byte, integer, real), the amount and quality of the data, the data compression, the instrument mode, the status of instrument peripheral hardware (filters, door positions, etc.), temperatures, spacecraft pointing, and space-craft mode, rate, and flare status.

The data index section can be expanded with a variety of structures to reflect data processing performed on the data saved in the file. A full history of the data processing can be maintained in this manner. The data described in the data index section immediately follows the data index section.

Optional data section. Some instruments require an additional data section to provide a complete copy of the telemetry data. The BCS spectral data comes down asynchronously, but there is other synchronous information available. For simplicity, a complete copy of the synchronous BCS data is available in the optional data section. For HXT, there are two one-dimensional arrays used to find the limb of the Sun. This data comes down at a different telemetry cadence and will be placed in the optional data section. SXT and WBS do not use the optional data section.

Road map section. The final section contains a short summary of each of the data sets available in the file and a pointer to the beginning of each data set. The section gives information on the instrument mode and a brief summary of the data (e.g., average or total counts). This section is almost an exact duplicate of the observing log entry, which is described in Section 4.

4. Log Files

A program will use the reformatted data files from the scientific instruments and create an observing log. This log will be produced automatically with little or no human interaction. Other log files will also be created using the reformatted data and the observing log as input. Some of these logs will require user interaction.

The observing log will provide a full summary of the modes and data available for each of the SOLAR-A instruments. There will be one entry for each SXT image, a BCS entry for each spectra (but not more often than every 4 s), and an entry every 2 major frames (every 4 s during high telemetry rate) summarizing the WBS and HXT count rates and modes. Information on the spacecraft orbital solutions, the reformatted data file IDs, and conversion coefficients are also provided in this log. A single file will contain the log for one month which should be about 30 Mbyte.

A user will be able to search the observing log and obtain a list of all cases where a given set of conditions are satisfied. For example, one could search for occasions when a certain mode occurs or when the signal was above a certain threshold. The user could cross reference to other instruments to select only cases where several instruments were in a given mode seeing a particular count level. It is possible that a similar log will eventually be generated with information on what ground-based instruments were in operation and the data available from that instrument. It will also be possible to create light curves from the instrument count rates with a time resolution of about 4 s.

An event log will be created listing the times when the instrument or spacecraft modes change. The event log will be generated from the observing log, but will be much smaller since it only logs changes in modes and will not contain information about the data count levels. Because of the small size, this log could easily contain several years of information and would remain readily transportable to machines with limited disk space.

A flare and active region log is contemplated. This log will require a user to identify which regions should be included. The log will probably contain information on which

M. D. MORRISON ET AL.

instruments saw that region at what times, and the mode that the instrument was in. Ground-based observations could also be incorporated into this log.

Acknowledgements

We thank I. Kondo for help in defining the SOLAR-A data base. The work at Lockheed was supported under contract NAS8–37334.